CHECK YOUR VOCABULARY FOR COMPUTING

a workbook for users

second edition

Original material by
David Riley

Edited and revised by
Liz Greasby

PETER COLLIN PUBLISHING

First published in Great Britain 1995
reprinted 1997
Second edition published 1999

Published by Peter Collin Publishing Ltd
1 Cambridge Road, Teddington, Middx, UK
© Peter Collin Publishing Ltd 1995, 1999

You may photocopy the inside pages of this workbook (pages numbered 1 to 64) for classroom use only and not for resale.

You are not licensed to copy the cover.
All rights reserved.

British Library Cataloguing in Publication Data
A catalogue record for this book is available from the British Library

ISBN 1-901659-28-3

Text typeset by PCP Ltd
Printed by Blackmore Limited, Longmead, Shaftesbury, Dorset.

Workbook Series
Check your:

Vocabulary for American Business	1-901659-32-1
Vocabulary for Banking and Finance	0-948549-96-3
Vocabulary for Business, 2nd edition	1-901659-27-5
Vocabulary for Colloquial English	0-948549-97-1
Vocabulary for Computing, 2nd edition	1-901659-28-3
Vocabulary for English	1-901659-11-9
Vocabulary for Hotels, Tourism, Catering	0-948549-75-0
Vocabulary for Law, 2nd edition	1-901659-21-6
Vocabulary for Medicine	0-948549-59-9

Specialist English Dictionaries

English Dictionary for Students	1-901659-06-2
Dictionary of Accounting	0-948549-27-0
Dictionary of Aeronautical English	1-901659-10-0
Dictionary of Agriculture, 2nd edition	0-948549-78-5
Dictionary of American Business, 2nd edition	1-901659-22-4
Dictionary of Automobile Engineering	0-948549-66-1
Dictionary of Banking & Finance, 2nd edition	1-901659-30-5
Dictionary of Business, 2nd edition	0-948549-51-3
Dictionary of Computing, 3rd edition	1-901659-04-6
Dictionary of Ecology & Environment, 3rd edition	0-948549-74-2
Dictionary of Government & Politics, 2nd edition	0-948549-89-0
Dictionary of Hotels, Tourism, Catering	0-948549-40-8
Dictionary of Human Resources, 2nd edition	0-948549-79-3
Dictionary of Information Technology, 2nd edition	0-948549-88-2
Dictionary of Law, 2nd edition	0-948549-33-5
Dictionary of Library & Information Management	0-948549-68-8
Dictionary of Marketing, 2nd edition	0-948549-73-4
Dictionary of Medicine, 2nd edition	0-948549-36-X
Dictionary of Military Terms	1-901659-24-0
Dictionary of Printing & Publishing, 2nd edition	0-948549-99-8
Dictionary of Science & Technology	0-948549-67-X

For details about our range of English and bilingual dictionaries and workbooks, please contact:
Peter Collin Publishing
1 Cambridge Road, Teddington, TW11 8DT, UK
tel: +44 181 943 3386   fax: +44 181 943 1673
email: info@pcp.co.uk   web site: www.pcp.co.uk

Introduction

The worksheets in this workbook contain a variety of exercises appropriate for students requiring a working knowledge of English computing terminology. The worksheets can be used either for self-study or in the classroom and can be completed in any order. Several have 'extensions': short classroom exercises based on the language in the main exercise. All the questions within this workbook are based on the *Dictionary of Computing* (ISBN 1-901659-04-6), also published by Peter Collin Publishing Ltd *(see the last page of this workbook for details of how request information or order this or any of our specialist dictionaries)*.

This workbook is aimed at students with at least an intermediate level of English. However, many people who work in computing have to read in English on a regular basis; students with a more basic level of English may therefore already have the passive vocabulary to handle many of the exercises.

Specialist vocabulary
It is important to appreciate that 'knowing' specialist vocabulary involves more than simply recognising it.

- You can understand the meaning of a word when reading or listening and yet be unable to remember that same word when speaking or writing.
- You may remember the word, but use it incorrectly. This can be a grammatical problem, like knowing that 'process' can be used both as a noun and as a verb. Or it may be a question of collocation: we use *machine* code, not *processor* code.
- Then there is the question of the sound of the word. Can you pronounce it? And do you recognise it when you hear it pronounced?

For these reasons - memory, use and sound - it is important that students practise specialist vocabulary so that they can learn to use it more confidently and effectively. The exercises in this workbook will help students to expand their knowledge and use of computing vocabulary.

Photocopiable material
All the worksheets can be legally photocopied to use in class. If, as a teacher, you intend to use most of the book with a class you may find it more convenient for the students to buy a copy each. You are not allowed to photocopy or reproduce the front or back cover.

Using the *Dictionary of Computing*
All of the vocabulary taught or practised in this workbook is in the Peter Collin Publishing *Dictionary of Computing*. The *Dictionary of Computing* gives definitions in simple English which students can read and understand. Many of the examples and definitions in the workbook are taken directly from the dictionary. Students should have a copy of the *Dictionary of Computing* for referring to when completing the exercises; using the dictionary is an essential part of successful language learning.

Structure of a *Dictionary of Computing* entry
Each entry within the dictionary includes key elements that help a student understand the definition of the term and how to use it in context. Each term has a clear example, and part of speech. This is followed by example sentences and quotations from newspapers and magazines that show how the term is used in real life. These elements of the dictionary are used to create the questions within this workbook.

Vocabulary Record Sheet
At the back of this book is a Vocabulary Record Sheet. Recording useful vocabulary in a methodical way plays a key role in language learning and could be done, for example, at the end of each lesson. The *Dictionary of Computing* is a useful tool for ensuring that the personal vocabulary record is accurate and is a good source for example sentences to show how words are used, as well as for notes about meaning and pronunciation, etc.

Workbook Contents

Page	Title	Description	Mode
	WORD-BUILDING		
1	Word association 1: missing links	Linking each set of four words with one other word	*Self-study*
2	Two-word expressions 1	Combining words from two lists to make two-word expressions that fit the definitions	*Self-study*
3	Word formation: nouns	Rewriting sentences using noun forms of verbs	*Self-study*
4	Two-word expressions 2	Combining words from two lists to make two-word expressions that fit the definitions	*Self-study*
5	Plural formation	Multiple choice: choosing correct plural forms of singular nouns	*Self-study*
6	Three-word expressions	Combining words from three lists to make two-word expressions that fit the definitions	*Self-study*
7	Word formation: adjectives	Rewriting sentences using adjective forms instead of nouns	*Self-study*
8	Opposites 1: prefixes	Selecting the correct prefix for each adjective to create an opposite; using the adjectives to complete sentences	*Self-study*
9	Word formation: verbs	Making verb forms from list of nouns; writing sentences using the verbs	*Self-study*
10	Word association 2: mind maps	Finding words in a mind map that fit definitions; designing mind maps	*Self-study*

Page	Title	Description	Mode
	PARTS OF SPEECH		
11	Nouns	Sentence completion	*Self-study*
12	Adjectives	Sentence completion	*Self-study*
13	Verbs 1	Sentence completion	*Self-study*
14	Verbs 2	Linking verbs with their definitions; sentence completion	*Self-study*
15	Verbs: past tense ~ regular verbs	Sentence completion	*Self-study*
16	Verbs: mixed tenses	Sentence completion	*Self-study*
17	Phrasal verbs 1	Linking phrasal verbs with their definitions	*Self-study*
18	Phrasal verbs 2	Sentence completion	*Self-study*
19	Verbs: active/passive	Changing sentences from active to passive	*Self-study*
20	Adverbs	Identifying adverbs in sentences and swapping adverbs around so that each sentence make sense	*Self-study*
21	Prepositions	Correcting sentences with deliberate mistakes in the prepositions	*Self-study*

Page	Title	Description	Mode
	PRONUNCIATION		
22	Word stress	Classifying three-syllable words by their pronunciation **Extension**: practising the sentences with a partner	*Self-study* *Pair work*
23	Present simple	Classifying verbs in present tense by pronunciation **Extension**: working with a partner to identify plural nouns in each pronunciation category	*Self-study* *Pair work*
24	Past simple	Classifying verbs in past tense by pronunciation	*Self-study*

Page	Title	Content	Mode
	VOCABULARY		
25	Good advice	Matching half-sentences together to make complete sentences **Extension**: writing pieces of good advice with a partner	*Self-study* *Pair work*
26	Odd one out	Identifying word that is different to others in set	*Self-study*
27	Opposites 2	Matching words with opposite meanings; inserting correct opposites in sentences **Extension**: working with a partner to test one another	*Self-study* *Pair work*
28	Abbreviations	Stating what abbreviations stand for **Extension**: working with a partner to test one another	*Self-study* *Pair work*
29	Telephone conversations	Placing sentences in correct order to make telephone conversations **Extension**: practising conversations	*Self-study* *Pair work*
30	Operating systems	Completing texts	*Self-study*
31	Instructions	Placing sentences in correct order to make sets of instructions; choosing title for each set **Extension**: giving instructions for a computing procedure	*Self-study*
32	Memory	Matching correct definitions and abbreviations with terms	*Self-study*
33	Internet	Completing text by inserting correct vowels	*Self-study*
34	This and that	Combining words from two lists to make expressions; using expressions to complete sentences	*Self-study*
35	Slang	Finding words in conversations that fit definitions	*Self-study*

Page	Title	Content	Mode
	PUZZLES & QUIZZES		
36 37	Communicative crossword 1	Completing crossword by working with partner and defining words	*Pair work*
38	Anagrams 1	Solving anagrams by reading clues and putting letters in order	*Self-study*
39	Word search	Finding words hidden in letters using clues listed	*Self-study*
40 41	Communicative crossword 2	Completing crossword by working with partner and defining words	*Pair work*
42 43	Communicative crossword 3	Completing crossword by working with partner and defining words	*Pair work*
44	Anagrams 2	Solving anagrams by reading clues and putting letters in order	*Self-study*
45	Computing crossword	Solving crossword	*Self-study*
46 47	Communicative crossword 4	Completing crossword by working with partner and defining words	*Pair work*
48 49	Communicative crossword 5	Completing crossword by working with partner and defining words	*Pair work*
50	Quiz	Answering questions **Extension**: writing a quiz with a partner	*Self-study* *Pair work*
51	Vocabulary Record Sheet	Recording new vocabulary, definitions and terms	

Page	Title	Content	Mode
	ANSWER KEY		
52	Answer key	Answers to all worksheets	

Using the workbook

Most students find it easier to assimilate new vocabulary if the words are learned in related groups, rather than in isolation. For example, words frequently occur in the same context as their opposites and, as such, it makes sense to learn the pairs of opposites together (*see worksheets on pages 8 and 27*). Similarly, mind maps encourage students to look for connections between words (*see worksheet on page 10*). The exercises and activities in this workbook have all been grouped into sections. These sections practise different elements of computing vocabulary, enabling the student to gain a fuller understanding of the words learnt.

The first section, **Word-building** (*pages 1-10*), encourages the student to identify links between words and to learn words that are morphologically related (for example, verbs and nouns which have the same stems). Within the **Parts of Speech** (*pages 11-21*) section, the emphasis is on understanding meanings and how to use terms in their correct grammatical forms. The worksheets in the third section practise the **Pronunciation** of computing vocabulary (*pages 22-24*). The section **Vocabulary in Context** (*pages 25-35*) includes topic-specific exercises such as completing texts on operating systems and the Internet. The activities in the last section, **Puzzles & Quizzes** (*pages 36-51*), expand students' knowledge and use of vocabulary in a fun way.

Communicative crosswords

Included in the last section are five communicative crosswords. These are speaking exercises where students complete a half-finished crossword by exchanging clues with a partner. There are two versions of the crossword: A & B. The words which are missing from A are in B, and vice versa. No clues are provided: the students' task is to invent them. This is an excellent exercise for developing linguistic resourcefulness; in having to define words themselves students practise both their computing vocabulary and the important skill of paraphrasing something when they do not know the word for it.

Using communicative crosswords

Stage 1 – Set-up. Divide the class into two groups - A and B - with up to four students in each group. Give out the crossword: sheet A to group A, sheet B to group B together with a copy of the dictionary. Go through the rules with them. Some answers may consist of more than one word.

Stage 2 - Preparation. The students discuss the words in their groups, exchanging information about the words they know and checking words they do not know in the ***Dictionary of Computing***. Circulate, helping with any problems. This is an important stage: some of the vocabulary in the crosswords is quite difficult.

Stage 3 - Activity. Put the students in pairs - one from group A and one from group B. The students help each other to complete the crosswords by giving each other clues.

Make sure students are aware that the idea is to help each other complete the crossword, rather than to produce obscure and difficult clues.

- What's one down?
- *It's a number of instructions that perform a particular task.*
- A program?
- *No, it's included as part of a program.*
- A routine?
- *Yes, that's right.*

A A	B B
A A	B B

Students work in groups, checking vocabulary.

Alternatively, students can work in small groups, each group consisting of two As and two Bs and using the following strategies:

i) defining the word
ii) describing what the item looks like
iii) stating what the item is used for
iv) describing the person's role
v) stating what the opposite of the word is
vi) giving examples
vii) leaving a gap in a sentence for the word
viii) stating what the word sounds like.

A B	A B
A B	A B

Students work in pairs, co-operating to solve their crosswords

Word association 1: missing links

Each of the sets of four words below can be linked by one other word. All the terms are relating to computing. What are the missing words? Write them in the centre of the charts. The first has been done for you as an example.

1. mechanical — *mouse* — pointer
   bus — *mouse* — driver

2. relational — __ — engine
   on-line — __ — language

3. floppy — __ — formatting
   backup — __ — drive

4. laser — __ — buffer
   ink-jet — __ — quality

5. touch — __ — saver
   text — __ — capture

6. data — __ — sharing
   disk — __ — transfer

Two-word expressions 1

Make 15 two-word expressions connected with computing by combining words from the two lists: A and B. Then match each expression with the appropriate phrase. Use each word once. The first one has been done for you as an example.

A	B
artificial	art
clip	circuit
desktop	database
read	disk
expanded	fibre
hard	friendly
information	intelligence
integrated	memory
electronic	multitasking
operating	mail
optical	~~processing~~
~~parallel~~	only
preemptive	publishing
relational	system
user	technology

1. Computer operating on several tasks simultaneously

   ......*parallel processing*..........................

2. Circuit where all the active and passive components are formed on one piece of semiconductor

   ....

3. Set of data where all items are related

   ....

4. Design, layout and printing of documents, books and magazines using special software

   ....

5. Computers that try to emulate human intelligence

   ....

6. Fine strand of glass or plastic used for the transmission of light signals

   ....

7. A feature of some operating systems that allows them to run several programs at the same time in an efficient manner

   ....

8. Software that is easy to use and interact with

   ....

9. Rigid magnetic disk that is able to store many times more data than a floppy disk and usually cannot be removed from the disk drive that is located inside a PC

   ....

10. Set of pre-drawn images or drawings that a user can incorporate into a presentation, report or desktop publishing document

    ....

11. Extra RAM fitted to an IBM PC-compatible that is located above the first 1Mb, but that needs a software driver before it can be used.

    ....

12. Way of sending and receiving messages between users on a network

    ....

13. Software that controls and co-ordinates the actions of the different parts of your computer

    ....

14. Technology involved in acquiring, storing, processing and distributing information electronically

    ....

15. File or memory device whose stored data cannot be changed

    ....

Word formation: nouns

A fast way to expand your vocabulary is to make sure you know the different forms of the words you learn. Rewrite the sentences below, changing the verbs (which are in **bold**) to nouns. Don't change the meaning of the sentences, but be prepared to make grammatical changes if necessary. For example:

*The two systems **interact***
*There's **interaction** between the two systems*

1. The transaction was **recorded** in the data base.
   There's ....
   ....

2. This system is easy to **install**.
   The ....
   ....

3. This screen **flickers** slightly.
   There's ....
   ....

4. The new PC will be **launched** in January.
   The ....
   ....

5. The system **failed** when I booted up this morning.
   There was ....
   ....

6. The factory is **equipped** for computer controlled production.
   The factory has ....
   ....

7. A maths co-processor **enhances** your system.
   A maths co-processor is ....
   ....

8. You'll have to **compare** the results of the two programs.
   You'll have to make ....
   ....

9. This is our system for **storing** client records.
   This is our ....
   ....

10. Only privileged users can **access** this information.
    Only privileged users have ....
    ....

11. It is sometimes possible to **recover** data from a corrupted disk.
    ....
    ....

12. The files are **retrieved** automatically.
    File ....
    ....

13. Jack is responsible for **maintaining** the system.
    Jack is responsible for ....
    ....

14. Something's wrong: the keyboard doesn't **respond**.
    Something's wrong: there's ....
    ....

Two-word expressions 2

Make 15 two-word expressions connected with computing by combining words from the two lists: A and B. Then match each expression with the appropriate phrase. Use each word once. The first one has been done for you as an example.

A	B
baud	analysis
catastrophic	bus
clean	~~code~~
device	degradation
flip	detector
floppy	directory
graceful	disk
interactive	driver
laser	error
local	flop
~~machine~~	machine
root	printer
speech	rate
systems	recognition
virus	video

1. Programming language consisting of commands in binary code that can be directly understood by the CPU without the need for translation

   ....*machine code*........................

2. Electronic circuit whose output can be one of two states, which can be used to store one bit of data

   ....

3. Secondary storage device

   ....

4. Computer that contains only the minimum ROM-based code to boot its system from disk

   ....

5. Analysing spoken words in such a way that they can be processed in a computer to recognise spoken words and commands

   ....

6. Starting node from which all paths branch in a data tree structure

   ....

7. Analysing a process to see if it could be carried out more efficiently by computer

   ....

8. Utility software which checks executable files to see if they have been infected with a known virus

   ....

9. High resolution output device

   ....

10. Direct link between a device and the processor

    ....

11. System that uses a computer linked to a mediadisc player to provide processing power and real image

    ....

12. Allowing some parts of the system to continue to function after a part has broken down

    ....

13. Error that causes a program to crash

    ....

14. Measure of number of signals transmitted per second

    ....

15. Routine used to interface and manage a peripheral

    ....

Plural formation

In *Column A* of this table there are 12 nouns relating to computing. For each of the nouns decide whether the correct plural form is in *Column B* or *Column C* and then circle it.

The first question has been done for you as an example.

	Column A	*Column B*	*Column C*
1.	virus	(viruses)	virii
2.	expansion card	expansion cards	expansions card
3.	appendix	appendixes	appendices
4.	key	keies	keys
5.	asterisk	asterisks	asterixes
6.	pixel	pixelae	pixels
7.	axis	axes	axises
8.	directory	directories	directorys
9.	criterion	criteria	criterions
10.	bureau	bureaus	bureaux
11.	formula	formulae	formulas
12.	font	fontes	fonts

Based on the **Dictionary of Computing**, third edition
ISBN 1-901659-04-6
Peter Collin Publishing Ltd

Three-word expressions

Word-building

Make 12 three-word expressions connected with computing by combining words from the three lists - A, B, C - and match each expression with the appropriate phrase. Use each word once. The first one has been done for you.

A	B	C
bulletin	access	example
~~central~~	area	exchange
dots	board	inch
dynamic	by	injury
graphical	character	interface
local	data	memory
near	down	menu
optical	letter	network
pull	per	quality
query	~~processing~~	recognition
random	strain	system
repetitive	user	~~unit~~

1. Control unit + arithmetic and logic unit + input/output unit

   .....*central processing..unit*....................

2. Pain in the arm felt by someone who performs the same movement many times over, as when operating a computer terminal

   ....

3. Interface between an operating system or program and the user

   ....

4. Memory that allows access to any location in any order

   ....

5. Information and message database accessible by modem and computer link

   ....

6. Set of options that are displayed below the relevant entry on a menu bar

   ....

7. Simple language used to retrieve information from a database management system

   ....

8. Network where various terminals and equipment are all a short distance from one another and can be interconnected by cables

   ....

9. Method by which two active programs can exchange data

   ....

10. Standard method used to describe the resolution capabilities of a page printer or scanner

    ....

11. Process that allows printed or written characters to be recognised optically and converted into machine-readable code that can be input into a computer

    ....

12. Printing by a dot-matrix printer that provides higher quality type, which is almost as good as a typewriter, by decreasing the spaces between dots

    ....

Based on the **Dictionary of Computing**, third edition
ISBN 1-901659-04-6
Peter Collin Publishing Ltd

Word formation: adjectives

The italicised words in the sentences in *Column A* are all nouns. What are the adjective forms? Complete the sentences in *Column B* using the correct adjective forms. The first question has been done for you as an example.

	Column A	*Column B*
1.	She asked about the IBM-*compatibility* of the hardware.	She asked whether the hardware was IBM-...*compatible.*
2.	The board has total *confidence* in the effectiveness of the new system.	The board is totally ...
3.	The *sophistication* of the new package is remarkable.	The new package is remarkably ...
4.	We checked the *validity* of the password.	We checked that the password was ...
5.	He commented on the *electroluminescence* of the TV screen coating.	He commented that the TV screen coating is ...
6.	The keyboarders are finding that the manuscript lacks *legibility*.	The keyboarders are finding that the manuscript is hardly ...
7.	What is the *difference* between these two products?	What makes these two products ...
8.	He is doubtful about the *efficiency* of the new networking system.	He is doubtful about whether the new networking system is ...
9.	We have the printer *capability* to produce high-quality colour images	Our printer is ...
10.	They reported that there was data *corruption* on the disk.	They reported that the data on the disk was ...

Based on the **Dictionary of Computing**, third edition
ISBN 1-901659-04-6
Peter Collin Publishing Ltd

Opposites 1: prefixes

Exercise 1. English often uses prefixes to create opposites. There are several different prefixes that are used. Choose the right prefix for each of the adjectives below and write them into the table. The first one has been done for you: a password that is *inaccurate* is not correct (accurate).

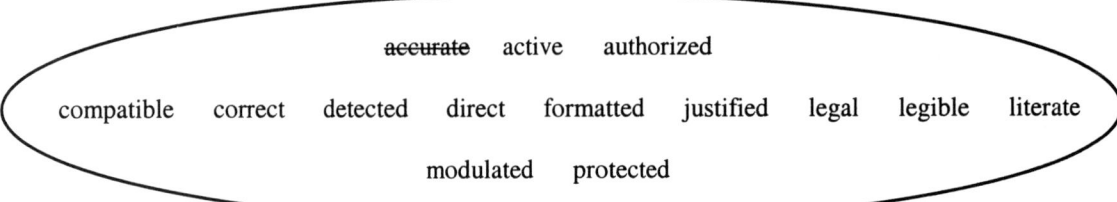

accurate active authorized compatible correct detected direct formatted justified legal legible literate modulated protected

il-	in-	un-
1.	1. *inaccurate*	1.
2.	2.	2.
3.	3.	3.
	4.	4.
	5.	5.
		6.

Exercise 2. Use eight of the adjectives in the table to complete these sentences. The first one has been done for you as an example.

1. It is impossible to copy to an ..*unformatted*.. disk.

2. The input data was ...., so the output is also incorrect.

3. They tried to link the two systems, but found they were ....

4. The programming error was .... for some time.

5. If the manuscript is ...., send it back to the author to have it typed.

6. There are still numerous people in this country who are ....

7. He entered an .... password.

8. Passwords are used to prevent .... access to the data.

Word formation: verbs

Exercise 1. The words listed in the table below are nouns. What are the verb forms of these nouns? The first question has been done for you as an example.

alteration .... *alter*	interaction
analysis	modification
assembly	multiplication
automation	prevention
communication	process
compilation	program
emulation	recovery
enhancement	removal
fluctuation	retrieval
generation	scan
installation	storage
instruction	use

Exercise 2. Choose ten verbs from Exercise 1 and write a sentence below for each one. Write the correct form of each verb in the column on the right and leave gaps for the verbs in the sentences. Cover up the right-hand column and give the sentences to another student as a test. For example:

These are the records .... in the search.	*retrieved*

1. ....
2. ....
3. ....
4. ....
5. ....
6. ....
7. ....
8. ....
9. ....
10. ....

Based on the **Dictionary of Computing**, third edition
ISBN 1-901659-04-6
Peter Collin Publishing Ltd

Word association 3: mind maps

A mind map is a way of organising vocabulary to show the connections between words. This mind map is based on the term 'desktop publishing'.

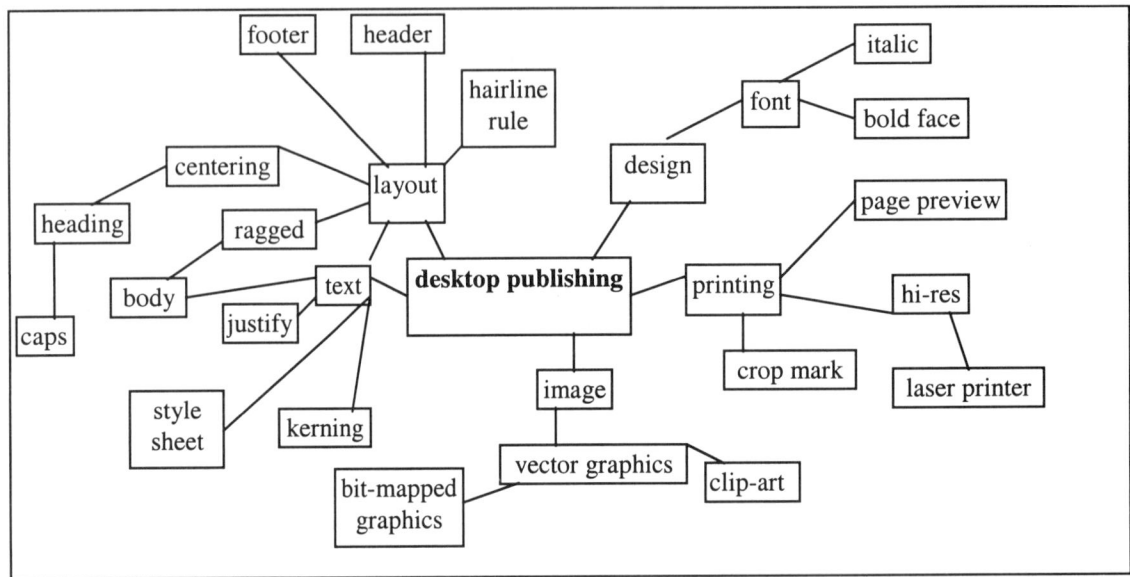

Exercise 1. Find words in the mind map that fit the following definitions.

1. Large form of letters as opposed to lower case ………………..
2. Template that can be preformatted to generate automatically the style or layout of a document ………………..
3. Set of pre-drawn images or drawings that a user can incorporate into a presentation or graphic ………………..
4. Set of characters all of the same style, size and typeface ………………..
5. Slight overlapping of certain printed character areas to prevent large spaces between them, giving a neater appearance ………………..
6. Main section of text in a document ………………..
7. Image whose individual pixels can be controlled by changing the value of its stored bit ………………..
8. Ability to display or detect a very large number of pixels per unit area ………………..
9. Copy of an original picture or design ………………..
10. Title or name of a document or file ………………..
11. Computer drawing system which uses line length and direction from an origin to plot lines ………………..
12. Graphical representation of how a page will look when printed, with different type styles, margins and graphics correctly displayed ………………..
13. Printed marks that show the edge of a page or image and allow it to be cut accurately ………………..
14. Very thin line ………………..

Exercise 2. Design a mind map for one or more of the following:
- database
- hardware
- Internet.

Nouns

There are 12 nouns connected with computing in the box below. Use them to complete the sentences. The first one has been done for you as an example.

connector	~~database~~	fault	field
model	modem	password	plaintext
platform	procedure	utility	virus

1. The sales department keeps the information about the company's clients in a ...*database*........ .

2. The user has to key in the .... before he can access the network.

3. The .... connects to one of the serial ports in your computer.

4. The .... at the end of the cable will fit any standard serial port.

5. The employee record has a .... for age.

6. This is the latest ....

7. This .... sorts all the files into alphabetical order.

8. A lost file cannot be found without a file-recovery ....

9. The technical staff are trying to correct a programming ....

10. The messages were sent as .... by telephone.

11. This software will only work on the IBM PC ....

12. If your PC is infected with a ...., your data is at risk.

Adjectives

Complete the sentences using the adjectives in the box. Use each adjective once only. The first one has been done for you as an example.

> clean   crash-protected   dedicated   downloadable   electroluminescent   excessive
> faulty   normal   ~~re-chargeable~~   redundant   common   unformatted   unpopulated
> user-friendly   concurrent

1. A ...*re-chargeable*.. battery is used for RAM back-up when the system is switched off.

2. I'll have to start again - I've just erased the only .... copy.

3. This is a .... fault with this printer model.

4. Each .... process has its own window.

5. There's only one .... graphics workstation in this network.

6. The .... procedure is for backup copies to be made at the end of each day's work.

7. It is impossible to copy an .... disk.

8. There must be a .... piece of equipment in the system.

9. If the disk is ...., you will never lose your data.

10. The screen coating is ....

11. The program used an .... amount of memory to accomplish the job.

12. The program is very ....

13. These fonts are ....

14. The parity bits on the received data are .... and can be removed.

15. You can buy an .... RAM card and fit your own RAM chips.

Based on the **Dictionary of Computing**, third edition
ISBN 1-901659-04-6
Peter Collin Publishing Ltd

Verbs

All the verbs in the box relate to computing matters. Use them to complete the sentences. The first question has been done for you as an example.

configure	disconnect	~~expand~~	generate	halt	install	paste	process	purge
	recover	run	save	simplify	simulate	undo		

1. If you want to hold so much data you will have to ...*expand*..... the disk capacity.

2. Hitting CTRL-S will ……………….. the program.

3. We will ……………….. the new data.

4. Each month, I ……………….. the disk of all the old email messages.

5. You've just deleted the paragraph, but you can ……………….. it from the option in the Edit menu.

6. You only have to ……………….. the PC once - when you first buy it.

7. Don't forget to ……………….. the file before switching off.

8. It is possible to ……………….. the data but it can take a long time.

9. We can ……………….. an image from digitally recorded data.

10. This software is able to ……………….. the action of an aircraft.

11. Function keys ……………….. program operation.

12. The new package will ……………….. on my PC.

13. The system is easy to ……………….. and simple to use.

14. Now that I have cut this paragraph from the end of the document, I can ……………….. it in here.

15. Do not forget to ……………….. the cable before moving the printer.

Verbs 2

Exercise 1. Link each verb on the left with its definition on the right. The first one has been done for you as an example.

Verb	Definition
1. assign	a. to write data to a location and, in doing so, to destroy any data already contained in that location
2. broadcast	b. to make part of a text stand out from the rest
3. transfer	c. to switch between two states
4. emulate	d. to start or run a program
5. highlight	e. to give a computer or someone a job of work
6. input	f. to copy or behave like something else
7. launch	g. to distribute information over a wide area or audience
8. monitor	h. to look after and supervise a process to make sure it is operating correctly
9. overwrite	i. to move data from one device or storage area to another
10. relay	j. to check or repair or maintain a system
11. service	k. to transfer data or information from outside a computer to its main memory
12. toggle	l. to receive data from one source and to retransmit it to another point

Exercise 2. Complete these sentences using the verbs from Exercise 1. You may have to make some changes to fit the grammar of the sentences. The first one has been done for you as an example.

1. All network signals are ...*relayed*.... to the next office using this controller.
2. The data was .... via a numeric keypad.
3. The 'bold' attribute can be .... on or off by pressing this function key.
4. The files were .... to the backup disk.
5. The disk drives were .... yesterday and are working well.
6. He .... the latest news over the WAN.
7. The machine .... each signal as it is sent out.
8. Some laser printers are able to .... the more popular office printers.
9. The latest data input has .... the old information.
10. Two PCs have been .... to outputting the labels.
11. The headings are .... in bold.
12. You .... the word processor by double clicking on this icon.

Based on the **Dictionary of Computing**, third edition
ISBN 1-901659-04-6
Peter Collin Publishing Ltd

Verbs: past tense ~ regular verbs

Use the past tense forms of the verbs in the box to complete the sentences. The first question has been done for you as an example.

> check   contain   crash   demonstrate
> design   disable   enter   establish   export   fail   print
> receive   update   ~~upgrade~~   view

1. We ...*upgraded*..... our printer.

2. She .... a new chip.

3. To prevent anyone changing the data he .... the keyboard.

4. The electricity was cut off and the computers ....

5. They .... which component was faulty.

6. The prototype disk drive .... its first test.

7. The user .... the pull-down menu by clicking on the menu bar at the top of the screen.

8. The computer .... data via the telephone line.

9. He .... his files.

10. We .... the data as a text file.

11. He .... the file management program.

12. They .... the computer printout against the invoices.

13. The printer we had previously .... at 60 characters per second.

14. The file that was lost .... important documents.

15. I .... the name on the list.

Based on the **Dictionary of Computing**, third edition
ISBN 1-901659-04-6
Peter Collin Publishing Ltd

Verbs: mixed tenses

Parts of Speech

All the verbs in the box relate to computing matters. Use them to complete the sentences. You may have to change the forms of the verbs to fit the grammar of the sentences. (Remember the five forms of English verbs - for example: take, takes, took, taken, taking.)

The first question has been done for you as an example.

```
access    cache    carry    computerize    conform
copy    degauss    extract    house    identify    lose
        modify    run    share    supersede
```

1. She ...*accessed*... the employee's file stored on the computer.

2. This CPU .... instructions so improves performance by 15 percent.

3. Our stock control has been completely ....

4. The software will not run if it does not .... to the operating system standards.

5. We can .... the files required for typesetting.

6. The maintenance engineers have .... the cause of the system failure.

7. Backing up involves .... current working files onto a separate storage disk.

8. All the current files were .... when the system crashed and we had no backup copies.

9. The keyboard was .... for European users.

10. Do not interrupt the spelling checker while it is ....

11. The facility is .... by several independent companies.

12. The new program .... the earlier one, and is much faster.

13. The fibre optic link .... all the data.

14. The r/w heads have to be .... each week to ensure optimum performance.

15. The magnetic tape is .... in a solid plastic case.

Phrasal verbs 1

Phrasal verbs are common in conversational English. They are made up of two words: a verb and a preposition. For example: This is the procedure to wake up the system. 'Wake up' means to start or initiate. Match each phrasal verb below with its correct definition. The first question has been done for you as an example.

Phrasal verb		Definition
1. back up	a.	to enter various identification data, such as a password, usually by means of a terminal to the central computer before accessing a program or data
2. break down	b.	to configure/initialize/define/start an application or system
3. fill up	c.	to allow a machine to stand idle for a time after switching on, to reach optimum operating conditions
4. key in	d.	to stop working because of mechanical failure
5. log off	e.	to enter a symbol or instruction at the end of a computing session to close all files and break the channel between the user's terminal and the main computer
6. log on	f.	to read data or a signal from a recording medium
7. play back	g.	to exit to the operating system, whilst the original application is still in memory and the user then returns to the application
8. plug in	h.	to make text continue without a break
9. power up	i.	to make a copy of a file or data or disk
10. run on	j.	to enter text or commands via a keyboard
11. set up	k.	to make something completely full
12. shell out	l.	to disconnect the power supply to a device
13. switch off	m.	to switch on or apply a voltage to an electrical device
14. warm up	n.	to make an electrical connection by pushing a plug into a socket

Based on the **Dictionary of Computing**, third edition
ISBN 1-901659-04-6
Peter Collin Publishing Ltd

Phrasal verbs 2

Each of the sentences below should contain a phrasal verb. Complete the sentences by choosing the correct verbs and prepositions from the two boxes. Be careful: sometimes you have to change the form of the verb. The first has been done for you as an example.

VERBS
back
break   call   key
log   log   ~~plug~~   round
shut   turn   turn
warm

PREPOSITIONS
down
down   down   ~~in~~
in   off   off   on   on
up   up   up

1. No wonder it isn't working: you haven't even ...*plugged*... it ...*in*...!

2. You have to give your password in order to .... to the system.

3. Pushing the big red button on the front .... the CPU ....

4. They .... the latest data.

5. If you don't .... regularly you could lose data.

6. I .... all the customers' addresses from the database and checked them on screen.

7. When we found the virus the first thing we did was to .... the entire system.

8. My computer's .... again! I need a new machine.

9. Don't forget to .... everything .... before you go home.

10. When you've downloaded the information you need, then .... from the system.

11. Your printout will arrive in a couple of minutes: the laser's still ....

12. The precise amount is 2.5341, but we can .... it .... to 2.5.

Based on the **Dictionary of Computing**, third edition
ISBN 1-901659-04-6
Peter Collin Publishing Ltd

Verbs: active/passive

Change the sentences below from active to passive tense. For example:

Active: This computer system uses a last in first out retrieval method.
Passive: A first in first out computer system is used by this computer system.

Remember that it is not always necessary to mention the *subject* in a passive sentence: For example:

Active: In this instance we overlooked the delay.
Passive: In this instance the delay was overlooked.

1. The compiler automatically corrected the syntax errors.

    Passive:

2. We used a balun to connect the coaxial cable to the twisted-pair circuit.

    Passive:

3. Base band local area networks support a maximum cable length of around 300m.

    Passive:

4. They calculated keyboarding costs on the basis of 5,500 keystrokes per hour.

    Passive:

5. We tried out the beta test software on as many different PCs as possible to try and find all the bugs.

    Passive:

6. This company is developing a new brand of screen cleaner.

    Passive:

7. The maintenance engineer found some defects in the equipment.

    Passive:

8. They traced the fault to a faulty cable.

    Passive:

9. That device controls the copy flow.

    Passive:

10. The operating system uses a metafile to hold data that defines where each file is stored on disk.

    Passive:

Adverbs

The sentences below do not read correctly. Identify the adverbs in the sentences and then swap the adverbs around so that each sentence makes sense.

Some of the adverbs could be used in several of the sentences, but in order to complete the exercise successfully, all the sentences must make sense.

1. We deal manually with the manufacturer, without using a wholesaler.

   ....

2. They were both alphabetically responsible for the successful launch of the new system.

   ....

3. The text is consecutively transmitted to an outside typesetter.

   ....

4. The files are arranged fully under the customer's name.

   ....

5. Processing time is electronically 10% lower than during the previous quarter.

   ....

6. The sections of the program run incorrectly.

   ....

7. A daisy wheel printer produces directly formed characters.

   ....

8. In spooling, the printer is acting equally of the keyboard.

   ....

9. The data was approximately keyboarded.

   ....

10. The paper has to be fed into the printer independently.

    ....

Prepositions

The sentences in this exercise contain **mistakes**. The mistakes are all in the prepositions and there are three types:

1. missing preposition   I spoke ^him about this last week. *to*
2. wrong preposition   We're meeting again ~~in~~ ^Tuesday. *on*
3. unnecessary preposition   I'll telephone ~~to~~ you tomorrow.

Find the mistakes and correct them.

1. The computer is a great aid to rapid processing in large amounts of data.

2. His background is of the computer industry.

3. Copy the files by the hard drive, C:, to the floppy drive, A:.

4. The cable has the wrong connector this printer.

5. The company is trying to improve the circulation information between departments.

6. The old data was contrasted at the latest information.

7. The smoke at the faulty machine quickly diffused through the building.

8. The user cannot gain access the confidential information in the file without a password.

9. The glare from the screen makes to my eyes hurt.

10. The company has been illegally copying at copyright software.

11. The software is manufactured in this country after licence.

12. We had a new phone system installed in last week.

Based on the **Dictionary of Computing**, third edition
ISBN 1-901659-04-6
Peter Collin Publishing Ltd

Pronunciation

Word stress

One of the keys to English pronunciation is *stress* - one syllable is emphasised more than the others. There are three possible patterns for three-syllable words:

A: Stress on the first syllable ■□□ For example: **per**-ma-nent
B: Stress on the second syllable □■□ For example: e-**lec**-tric
C: Stress on the third syllable □□■ For example: in-ter-**rupt**

Read the conversations below and find all the three-syllable words. Underline them and classify them in the groups on the right. The first one has been done for you as an example.

Conversation 1
- Do you know how to operate this word processor?
- ○ A little bit. What do you want to know?
- How do I put a word in italics?
- ○ *Position the cursor at the beginning of the word, highlight the word like this and then click on the italics icon.*

Conversation 2
- This is my new portable computer.
- ○ *Very nice. How much memory has it got?*
- ○ Sixty-four megabytes of RAM and 4.6-gig on the hard disk.
- *So, you'll be able to work from home now.*

Conversation 3
- Windows is a graphical user interface.
- ○ *What does that mean?*
- Well, it's an easy way to interact with your programs.
- ○ *And how does it work?*
- It uses graphics or icons to represent functions or files and to allow the user to control software more easily.

Conversation 4
- What will change when we introduce the new system?
- ○ First, all transactions will be recorded in one place.
- That sounds more efficient.
- ○ And you'll be able to generate statistics any time you want.
- That's good.

Group A: ■□□
1 operate
2
3
4
5
6
7
8
9

Group B: □■□
1
2
3
4
5
6
7
8

Group C: □□■
1
2
3

Extension. Practise the conversations with a partner.

Present simple

Verbs in the present tense add an 's' in the third person singular: I work, you work, he/she/it works. But the 's' has three different pronunciations. Look at these examples:

A: /s/, for example *works*
B: /z/, for example *sells*
C: /ɪz/, for example *closes*

Find the third person present tense verbs in these sentences and classify them by their pronunciation. Put them in the correct columns in the table on the right. Be careful: some sentences have more than one example. There are 23 verbs in total. The first one has been done for you as an example.

A: /s/
_____

1. John browses on the Internet and downloads anything that looks interesting.
2. The company designs high specification workstations.
3. I've set up your computer so that it automatically boots up in Windows™.
4. The team programs in several different languages.
5. The software sends and receives mail from within any word-processing program, and prints it out on your laser printer.
6. This word-processing program corrects common errors as you type.
7. This instruction reads the first record of a file.
8. The program accesses the information on the hard disk and outputs it to the screen.
9. This utility detects and eliminates most viruses.
10. If you buy a modem make sure it conforms to Hayes™ standards.
11. Call me on this number if your machine crashes again.
12. When the user installs this program it automatically checks the specifications of the PC and adapts to them.
13. On the first of every month the program updates the list and faxes it to all members of the group.
14. Using this procedure ensures that unauthorised users cannot enter the system.

B: /z/
_____

C: /ɪz/
_____
browses

Extension. The same rule applies to plural nouns: /s/ chips, /z/ bugs, /ɪz/ devices. Work with a partner and find five nouns in each pronunciation category.

Pronunciation

Past simple

Regular verbs have three different pronunciations in the past tense (or the past participle). The difference is in the sound you use for the ending. Look at these examples:

A: /t/, for example *click__ed__*
B: /d/, for example *dragg__ed__*
C: /ɪd/, for example *point__ed__*

Find the past tense verbs in these sentences and classify them by their pronunciation. Put them in the correct columns in the table on the right. Be careful: some sentences have more than one verb. There are 25 examples in total. The first one has been done for you as an example.

1. The computer <u>received</u> data via the telephone line.
2. Errors were introduced into the text at keyboarding.
3. I downloaded information from the Internet about this subject.
4. I found the problem when I launched this program.
5. I selected the network laser, printed the document and closed down.
6. I typed the password and opened the file.
7. I warned my boss that there was going to be a problem.
8. Oh no! I've deleted all the client records!
9. The new version of this software was released in July.
10. The printout's fine: I checked it.
11. These machines haven't been serviced for a year.
12. They programmed the machine to find the shortest possible route between sales calls.
13. We dumped all the information onto the hard disk of the server.
14. We moved the DP department to the third floor.
15. We searched the database but your transaction wasn't recorded.
16. We've networked all the machines on the fourth floor.
17. With this system we've eliminated all possibilities of error.
18. You've saved this file in the wrong directory.
19. He reconfigured the field structure in the file.
20. I highlighted the headings in bold.

A: /t/

B: /d/
received

C: /ɪd/

Based on the **Dictionary of Computing**, third edition
ISBN 1-901659-04-6
Peter Collin Publishing Ltd

Good advice

These sentences all give advice, but they have been divided into separate halves. Match the half-sentences in *Column A* with the half-sentences in *Column B* to make 16 logical sentences.

For example: If you want to learn to keyboard,…you should use every opportunity to practise.

Column A	*Column B*
1. If you want to reduce screen flicker, …	a) …add a maths co-processor.
2. If you want to produce complicated graphics, …	b) …buy a laptop.
3. If you want to store more information, …	c) …buy a laser printer.
4. If you want to speed up your computer, …	d) …fit a faster microprocessor.
5. If you want to do CAD, …	e) …fit a modem.
6. If you want to share information through the company, …	f) …fit a sound card.
7. If you want to reduce the noise level in the office, …	g) …get a bigger hard disk.
8. If you want to send faxes from your PC,	h) …get a CD-ROM drive.
9. If you want to stop the computer when it hangs, …	i) …get an Apple Mac with a 21" screen.
10. If you want to import graphics from paper, …	j) …get a scanner.
11. If you want to reduce typing time, …	k) …hit Control, Alt & Delete.
12. If you want to make music on your PC, …	l) …make a backup.
13. If you want to do a lot of DTP, …	m) …network the computers.
14. If you want to take your work away with you, …	n) …put more memory in the printer
15. If you want to use interactive software, …	o) …use a non-interlaced monitor
16. If you want to protect your data, …	p) …use an OCR program.

Extension. Working with a partner, write five more sentences giving advice about computers.

Odd one out

In each set of words below, one term is the odd one out: different from the others. Find the term that is different, and circle it. For example:

monitor .... printer .... scanner .... (spreadsheet)

Spreadsheet is the odd one out. A spreadsheet is an application, the other terms are peripherals.

1. desktop .... laptop .... notebook .... palmtop
2. keyboard .... modem .... mouse .... trackball
3. compact .... floppy .... hard .... soft
4. bps .... dpi .... MIPS .... ppm
5. drive .... motherboard .... port .... power switch
6. local bus .... microprocessor .... graphics card .... port
7. model .... joystick .... parallel .... serial
8. keyboard .... modem .... monitor .... printer
9. database .... file .... spreadsheet .... word processor
10. client .... peer .... server .... standalone
11. RAM .... cache .... flash .... printout
12. cell .... column .... row .... window
13. click .... italics .... bold face .... caps
14. function key .... screen .... shift .... space bar
15. bug .... error .... howler .... message
16. allocate .... emulate .... replicate .... simulate
17. click .... drag .... point .... type
18. LCD .... QBE .... TFT .... VGA
19. inkjet .... laser .... plotter .... scanner
20. beep .... bloop .... wipe .... zap

Based on the **Dictionary of Computing**, third edition
ISBN 1-901659-04-6
Peter Collin Publishing Ltd

Opposites 2

Exercise 1. Find the words in list B which are opposite in meaning to the italicized words in list A. For example: *The opposite of turn on is turn off.*

A	B
1. The opposite of *anode* is	built-in
2. The opposite of *authorize* is	cancel
3. The opposite of *automatic* is	cathode
4. The opposite of *backward* is	close
5. The opposite of *boot up* is	complicated
6. The opposite of *character based* is	divide
7. The opposite of *column* is	duplex
8. The opposite of *confirm* is	forbid
9. The opposite of *continue* is	forward
10 The opposite of *delete* is	graphical
11 The opposite of *flexible* is	interrupt
12 The opposite of *hardware* is	manual
13 The opposite of *landscape* is	multiple
14 The opposite of *multiply* is	portrait
15 The opposite of *open* is	receive
16 The opposite of *add-on* is	restore
17 The opposite of *simplex* is	rigid
18 The opposite of *simple* is	row
19 The opposite of *single* is	shut down
20 The opposite of *transmit* is	software

Exercise 2. Complete these sentences using words from Exercise 1.

1. The communications ……………….. will only work with Hayes™-compatible modems.

2. ……………….. error correction is an error detection and correction method that is applied to received data to correct errors rather than requesting another transmission.

3. The computer has a ……………….. hard disk.

4. Put the total at the bottom of the ……………….. .

5. The systems manager has to ……………….. the purchase of a new computer.

Extension. Work with a partner and test each other. One partner closes the book, while the other asks questions such as: *What's the opposite of 'simplex'?*

Based on the **Dictionary of Computing**, third edition
ISBN 1-901659-04-6
Peter Collin Publishing Ltd

Vocabulary

Abbreviations

All these abbreviations are connected with computing. How many of them do you know? Check the ones you don't know in the dictionary. Write the full expressions on the right The first one has been done for you as an example.

1. BBS — *Bulletin Board System*
2. BIOS — ....
3. bps — ....
4. CAD — ....
5. DP — ....
6. dpi — ....
7. FAT — ....
8. HMA — ....
9. IKBS — ....
10. ISA — ....
11. IT — ....
12. LAN — ....
13. LCD — ....
14. MIPS — ....
15. OCR — ....
16. OS — ....
17. PDA — ....
18. QBE — ....
19. RISC — ....
20. TSR — ....
21. WAN — ....
22. WIMP — ....
23. WORM — ....
24. WP — ....
25. WYSIWYG — ....

Extension. Work with a partner and test each other. One partner closes the book, while the other asks questions such as *'What does BBS stand for?'*

Based on the **Dictionary of Computing**, third edition
ISBN 1-901659-04-6
Peter Collin Publishing Ltd

Telephone conversations

The lines in these telephone conversations are in the wrong order. Work out the correct order and write the sequence in the boxes. The first line in each conversation has been marked for you as an example.

Conversation A.

	£90. Would you like to order it now?
	Certainly. Which model do you have?
	64 megabytes.
	Goodbye.
	And how much memory have you got in it at the moment?
	No. No, I'd like to think about it. Thank you for the information.
	Not at all. Goodbye.
1	PC Memory Mart. Can I help you?
	That sounds alright. How much is it?
	The 333S.
	There's a 32 meg module - that's an upgrade to 96.
	Yes. Could you could give me some information about memory for Zell computers?

Conversation B.

	Do you get an error message when you try?
	Goodbye.
	I can't boot up the system.
	I see. Can you give me your name and number? I'll get a technician to call you.
	I'm sorry to hear that. What kind of problem is it?
	No - it just goes down a minute or two after starting.
	Not at all. Goodbye.
	Someone will call you within an hour, Mr Dent.
	Thank you.
1	Viking Computers. Can I help you?
	Yes. I'm Arthur Dent, D-E-N-T, and my number's 223 9898.
	Yes. I've just bought one of your machines and I've got a problem.

Extension. Practise the conversations with a partner.

Operating systems

The texts below relate to specific operating systems. Complete each of the four texts, using the words in the boxes above them. The first gap has been filled for you as an example.

Text 1. MS-DOS™

| control | functions | MS-DOS |
| software | version | Windows |

MS-DOS™ is operating system ....*software*.... developed by Microsoft that controls and co-ordinates the basic .... of your computer. If you are using Windows 95 or a later .... of Windows, the functions of MS-DOS have been integrated. If you are using Windows 3.1x or do not have ...., then you are relying on .... (or a similar product from IBM called PC-DOS) to .... the computer.

Text 2. Windows™

| commands | icons | Microsoft |
| mouse | multitasking | user |

Windows™ is a .... graphical .... interface for the IBM PC developed by .... Corp. that is designed to be easy to use. Windows™ uses .... to represent files and devices and can be controlled using a ...., unlike MS-DOS which requires .... to be typed in.

Text 3. Windows 95™

| filenames | interface | Internet |
| memory | networks | processor |

Windows 95™ provides support for long ...., an .... that's easier to use and better support for .... and the .... It does, however, require a faster .... and more .... to get good results - an absolute minimum of 8Mb and a fast 80486 are required.

Text 4. Windows 98™

| communications | configure | enhanced |
| features | version | |

Windows 98™ is an .... of Microsoft's Windows 95™ that provides more .... and internet .... and is easier to use and ....

30 Based on the **Dictionary of Computing**, third edition
ISBN 1-901659-04-6
Peter Collin Publishing Ltd

Instructions

The lines in these sets of instructions are in the wrong order. Put each set of instructions in the most logical order, and then choose a title for it. The first line in each instruction set has been marked for you and there are six lines in each instruction set.

Instruction Set A: *title*..

	Click on File in the menu bar and select Format from the pull-down menu.
	Double-click on the My Computer icon.
1	Turn on the computer using the power switch on the front.
	Turn on the monitor.
	Click on the 3 $1/2$ Floppy (A): icon.
	Put a new disk in the drive.

Instruction Set B: *title*..

	Select Print from the pull-down menu.
	Click on OK in the dialogue box.
	Type your text.
	Collect your printout from the printer.
1	Click on Start, Programs and Microsoft Word™.
	Click on File in the menu bar.

Instruction Set C: *title*..

	Disconnect the printer from the power supply and unplug the parallel interface cable.
	Open the panel at the back.
	Push the new memory board in and replace the panel.
	Reconnect the printer to the PC and to the power supply.
	Slide the old memory board out of the slot.
1	Turn off the computer and printer.

Extension. Give instructions for a procedure you use in computing.

Memory

Vocabulary

The terms below all relate to memory - the storage space in a computer system or medium that is capable of retaining data or instructions. Match the correct abbreviations and definitions with each term. Write the solutions in the right-hand column. The first one has been done for you as an example.

Term	Abbreviation	Definition	Solution
(1) Dynamic Random Access Memory	(i) SAM	(a) memory device that has had data written into it at the time of manufacture, and now its contents can only be read	*(1), (iii), (e)*
(2) Erasable Programmable Read-Only Memory	(ii) MMU	(b) system in a PC that defines extra memory added above the 640Kb of conventional memory	
(3) Expanded Memory System	(iii) DRAM	(c) register in a CPU that temporarily buffers all inputs and outputs	
(4) Memory Buffer Register	(iv) ROM	(d) electronic component in a computer that manages the way in which data is stored in different RAM chips	
(5) Memory Management Unit	(v) EMS	(e) memory components that will retain information as long as they are supplied with an electric current	
(6) Random Access Memory	(vi) EPROM	(f) storage where a particular data item can only be accessed by reading all the previous items in the list	
(7) Read Only Memory	(vii) MBR	(g) memory that allows access to any location in any order, without having to access the rest first	
(8) Serial-Access Memory	(viii) RAM	(h) component that can be programmed using a special electrical signal and will retain this information even without electrical power; usually erasable with ultraviolet light	

Based on the **Dictionary of Computing**, third edition
ISBN 1-901659-04-6
Peter Collin Publishing Ltd

Internet

All the vowels (A, E I, O, U) have been removed from this text on the internet and replaced with asterisks (*). Can you read it?

Th* *nt*rn*t *s *n *nt*rn*t**n*l n*tw*rk th*t l*nks t*g*th*r th**s*nds *f c*mp*t*rs *s*ng t*l*ph*n*s *nd c*bl* l*nks; th*s* c*mp*t*rs *r* c*ll*d th* s*rv*rs *nd *r* r*th*r l*k* * l*c*l t*l*ph*n* *xch*ng* - *nd*v*d**l *s*rs c*n th*n *s* * m*d*m t* c*nn*ct t* th* s*rv*r c*mp*t*r *nd s* h*v* *cc*ss t* th* *nt*r* w*rld n*tw*rk. * *s*r c*n s*nd *l*ctr*n*c m**l *v*r th* *nt*rn*t *nd tr*nsf*r f*l*s *nd t*xt fr*m *n* c*mp*t*r *n L*nd*n t* *n*th*r *n N*w Y*rk - *ll f*r th* pr*c* *f * l*c*l ph*n* c*ll t* th* n**r*st s*rv*r.

Th* W*rld W*d* W*b *s *n *nh*nc*m*nt t* th* *nt*rn*t *nd pr*v*d*s * gr*ph*c*l fr*nt-*nd t* th* d*ff*r*nt d*t*b*s*s *nd s*rv*rs th*t *r* *v**l*bl*. *n *rd*r t* c*nn*ct t* th* *nt*rn*t, y**'ll n**d * m*d*m *nd *n *cc**nt w*th * s*rv*r - n*rm*lly c*ll*d *n *nt*rn*t s*rv*c* pr*v*d*r (*SP) *r p**nt-*f pr*s*nc* pr*v*d*r - t*g*th*r w*th s*m* sp*c**l s*ftw*r*. Y**'ll b* g*v*n * *n*q** *D n*m* th*t w*ll (l*k* y**r t*l*ph*n* n*mb*r) *d*nt*fy y** t* *ny *th*r *s*r *n th* w*rld, t*g*th*r w*th *n *cc*ss t*l*ph*n* n*mb*r th*t *ll*ws y**r m*d*m t* c*nn*ct t* th**r s*rv*r. N* *n* p*rs*n *r c*mp*ny c*ntr*ls th* *nt*rn*t.

Based on the **Dictionary of Computing**, third edition
ISBN 1-901659-04-6
Peter Collin Publishing Ltd

Vocabulary

This and that

Use the words in the box to make eleven expressions connected with computing. Then use the expressions to complete the sentences. All the expressions follow the same pattern: x and y. The first one has been done for you as an example.

A		B
~~bells~~	&	click
cut		columns
drag		drop
hyphenation		embedding
point		flutter
rows		justification
search		paste
terminate		replace
tilt		stay resident
object linking		swivel
wow		~~whistles~~

1. This is just a basic program - it doesn't have any ... **bells and whistles** ....

2. If you change your mind you can use .... to change all the examples in the text.

3. If you want to add a comment to your information in your report you can use .... to get information from the word processor and copy it into the worksheet.

4. The speakers on that PC are very cheap - listen to the amount of .... they have!

5. You use a mouse to navigate a GUI: you can simply .... on icons to make most selections.

6. I use a little .... program to check for viruses.

7. If you .... the document icon onto the word processor icon, the system will start the program and load the document.

8. An American .... program will not work with British spellings.

9. Information in a spreadsheet is organised in ... .

10. Window uses .... to share data between applications.

11. We use........................... monitors for ergonomic reasons.

Based on the **Dictionary of Computing**, third edition
ISBN 1-901659-04-6
Peter Collin Publishing Ltd

Slang

In the world of computers there is special slang, just as in any other profession. The ten conversations below each contain an example of computer slang. Find the slang words and match them to the definitions on the right.

i How's your computer? Is it working now?
   It seems alright - it passed the smoke test.

ii Have you finished with this file?
    Yes - go ahead and zap it.

iii Can you fix this for me?
     I'm a bit short of time. I can do a kludge for you and have another look tomorrow.

iv What's wrong?
    I don't know. Some sort of gremlin - the system keeps going down.

v I'm having problems running Windows™.
   I'm not surprised. You need more than a couple of megs to do multitasking.

vi What did you do before you worked here?
    I was a project manager for Big Blue.

vii They say they're releasing a new version soon.
     Don't get excited. They've already made four product announcements. It's just vapourware.

viii What do you think the problem was?
      I don't know, but it seems alright now. It was probably just a glitch.

ix What kind of standard has been used for the network?
    Cheapernet - it's less expensive than Thick-wire Ethernet.

x What are you working on this week?
   I'm doing a comms program to get information from branch offices more easily.

Find the words which mean:

1. Communications

....

2. IBM

....

3. An unexplained fault in a system

....

4. Products which exist in name only

....

5. To wipe off all data currently in the workspace

....

6. Slang for a test which indicates that the machine must be working if no smoke appears when it is switched on

....

7. Megabytes

....

8. A temporary correction

....

9. Anything which causes a sudden, unexpected failure of a computer or equipment

....

10. Thin-wire Ethernet

....

Based on the **Dictionary of Computing**, third edition
ISBN 1-901659-04-6
Peter Collin Publishing Ltd

Communicative crossword 1 — sheet A

This crossword is not complete: you have only half the words. The other half are on sheet B. Check that you know the words in your crossword. Then work with a partner who has sheet B to complete the two crosswords. Follow these three rules:

1. Speak only in English.
2. Don't say the word in the crossword.
3. Don't show your crossword to your partner.

"What's 1 across?"

→ across, ↓ down

¹K		²					³A	⁴S	Y	N	⁵C
E	■		■		■	⁶	■		■		D
R	■		⁷T			■	⁸H	U	B	■
N	■		R		■		■		■	⁹A
¹⁰E	N	C	L	O	S	E	■	¹¹			P
L	■		J		■		■		■		P
■		¹²	A			■		■		■	L
¹³	■	¹⁴	■	N		■	¹⁵			■	I
¹⁶				H	■		■		■		C
	■		¹⁷O	K		¹⁸	■		■		A
¹⁹			R	■			■		■		T
	■		²⁰S				■	²¹B	■	I
²²T	A	²³B	L	E			■	U	■	O
■		A	■		²⁴		■	N	■	N
²⁵B	A	C	K	²⁶G	R	O	U	²⁷N	D	■
O	■	K	■	A	■		²⁸S	L	O	²⁹T
O	■	U	■	M	■		■	E	■	A
³⁰T	O	P	■	³¹E			■	D	■	B

36   Based on the **Dictionary of Computing**, third edition
ISBN 1-901659-04-6
Peter Collin Publishing Ltd

Communicative crossword 1   sheet B

This crossword is not complete: you have only half the words. The other half are on sheet A. Check that you know the words in your crossword. Then work with a partner who has sheet A to complete the two crosswords. Follow these three rules:

1. Speak only in English.
2. Don't say the word in the crossword.
3. Don't show your crossword to your partner.

"What's 1 across?"

→ across, ↓ down

1 K	L	2 U	D	G	E	■	3	4 S	■	■	5
	■	N	■	■	■	6 M	■	C	■	■	
	■	L	■	7	■	O	■	8 H	■	■	■
	■	O	■		■	V	■	E	■	■	9
10		C	■		■	E	■	11 D	U	M	P
	■	K	■	■	■	M	■	U	■	■	■
■	■	■	12 D	A	T	E	■	L	■	■	■
13 A	■	14 F	■	■	■	N	■	15 E	M	■	■
16 F	L	A	S	H	■	T	■	R	■	■	■
F	■	S	■	17		■	18 T	■	■	■	■
19 E	N	T	E	R	■	■	Y	■	■	■	■
C	■	■	■	20 S	K	I	P	■	21	■	■
22 T	■	23	■	■	■	■	E	■		■	■
■	■	■	■	■	■	24 P	■	■	■	■	■
25		■		26	■	O	■	27		■	■
■	■		■	■	■	I	■	28		■	29
■	■		■	■	■	N	■		■	■	
30		■	■	31 E	X	T	E	N	D	■	

Based on the **Dictionary of Computing**, third edition
ISBN 1-901659-04-6
Peter Collin Publishing Ltd

Anagrams 1

Solve the anagrams by reading the clues and putting the letters in order to form words. Enter the solutions in the table to find the mystery word. The first one has been done for you as an example.

1. Waiting to be used .... ADERY
2. List of data in columns and rows .... ABELT
3. Where one system ends and another begins .... ACEEIFNRT
4. Method of organising files stored on disk .... CDEIORRTY
5. System of words or symbols which allows communication with computers .... AAEGGLNU
6. Data which is out of date or which contains errors .... AABEGGR
7. To carry out or to put something into action .... EEILMMNPT
8. "The user has to ... herself to the system by using a password" .... DEFIINTY
9. With no errors .... ACELN
10. Taking place at the same time .... AEILMNOSSTUU
11. A number of separate moving parts or components acting together
    to carry out a process .... ACEHIMN
12. To modify a system to a customer's requirements .... CEIMOSTUZ
13. Byte made up of five bits .... EINQTTU
14. Cannot be anticipated .... ADMNOR
15. To set up a new computer system to the user's requirements .... SLANTIL

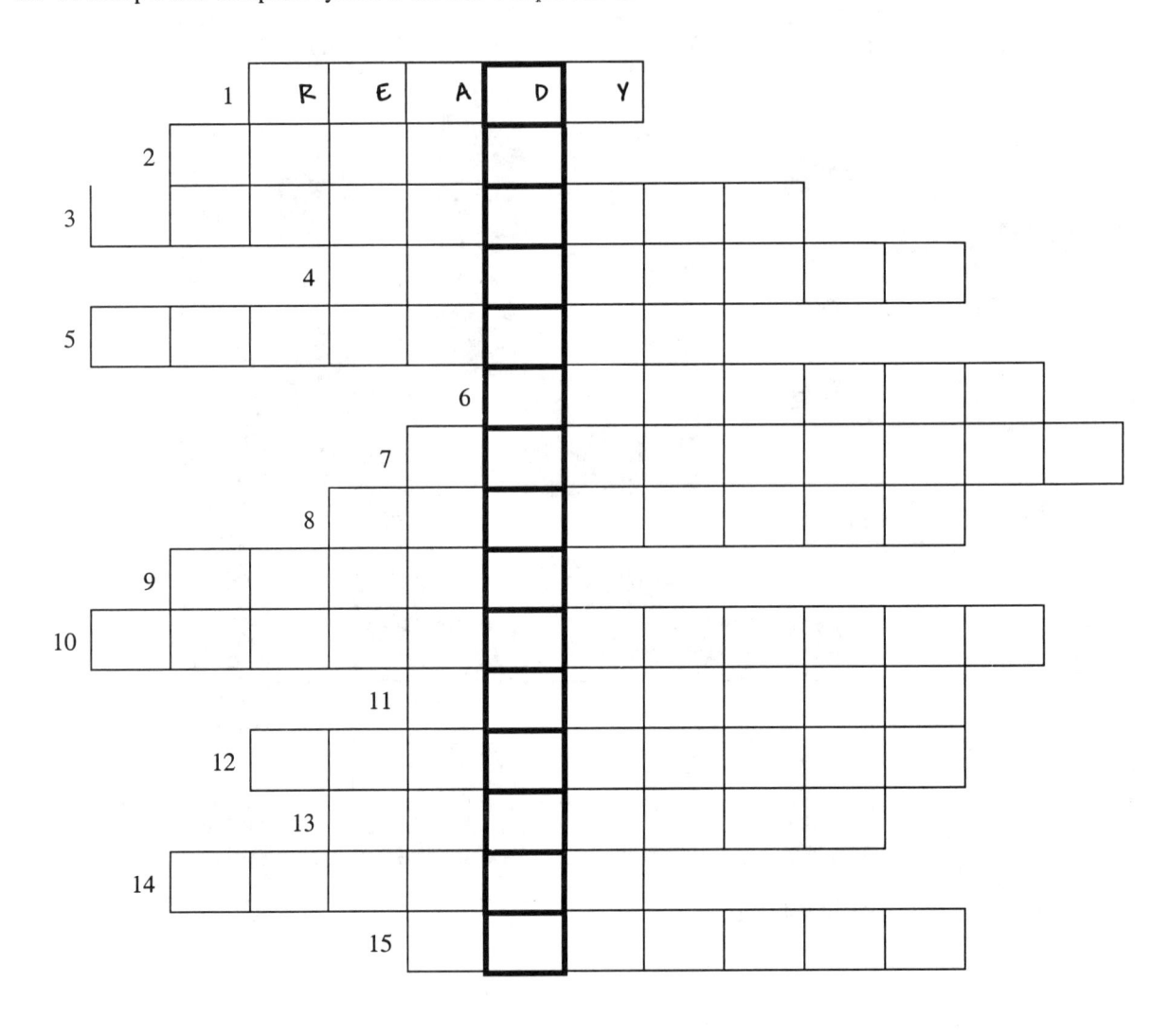

38   Based on the **Dictionary of Computing**, third edition
ISBN 1-901659-04-6
Peter Collin Publishing Ltd

Word search

Find the 24 computing terms and expressions hidden in the letters below; 14 read across and 10 read down. The first word has been found for you as an example. The clues listed beneath will help you to find all of the words.

~~M~~	~~U~~	~~L~~	~~T~~	~~I~~	~~M~~	~~E~~	~~D~~	~~I~~	~~A~~	A	T
A	B	W	O	R	K	S	H	E	E	T	E
I	N	V	A	L	I	D	C	F	L	E	X
L	O	W	O	P	E	R	A	T	O	R	T
B	D	I	D	E	C	F	G	H	I	M	J
O	E	P	K	P	O	R	T	I	L	I	M
X	N	E	R	O	N	P	Q	C	R	N	S
S	V	I	E	W	T	G	L	O	B	A	L
A	L	I	S	T	R	T	U	N	V	L	O
V	W	D	E	F	A	U	L	T	X	Y	A
E	A	R	T	H	S	E	A	R	C	H	D
P	I	X	E	L	T	S	U	R	G	E	Z

1. Difference between black and white or between colours
2. Predefined course of action or value that is assumed unless the operator alters it
3. To connect an electrical device to the earth
4. Wire or cable used to connect an appliance to the mains electricity supply
5. Covering everything
6. Graphic symbol or small picture displayed on screen used in an interactive computer system to provide an easy way of identifying a function
7. Not valid
8. Series of ordered items of data
9. To put a disk or tape into a computer, so that it can be run
10. Electronic storage space with an address in which a user's incoming messages are stored
11. The combination of sound, graphics, animation, video and text within an application
12. Person who makes a machine or process work
13. Interconnection point in a structure or network
14. Smallest single unit or point of a display whose colour or brightness can be displayed
15. Socket or physical connection allowing data transfer between a computer's internal communications channel and another external device
16. To return a system to its initial state, to allow a program or process to be started again
17. To store data or a program on an auxiliary storage device
18. Process of looking for and identifying a character or word or section of data in a document or file
19. Sudden increase in electrical power in a system, due to a fault or noise or component failure
20. Device usually made up of a display unit and a keyboard which allows entry and display of information when on-line to a central computing system
21. Alphanumeric characters that convey information
22. To look at something, especially something displayed on a screen
23. To clean data from a disk
24. (In a spreadsheet program) a two-dimensional matrix of rows and columns that contain cells which can, themselves, contain equations

Based on the **Dictionary of Computing**, third edition
ISBN 1-901659-04-6
Peter Collin Publishing Ltd

Communicative crossword 2 — sheet A

Puzzles & Quizzes

This crossword is not complete: you have only half the words. The other half are on sheet B. Check that you know the words in your crossword. Then work with a partner who has sheet B to complete the two crosswords. Follow these three rules:

1. Speak only in English.
2. Don't say the word in the crossword.
3. Don't show your crossword to your partner.

"What's 1 across?"

→ across, ↓ down

¹D		²		³C		⁴O						
I				E		X		⁵		⁶		
⁷S	E	M	A	N	T	I	C	S				
T				T		D		⁸A				
R				R		⁹E	N	¹⁰D	I	N	G	
¹¹I			¹²	E				E				
B						¹³T	I	F	F			
¹⁴U	S	A	S	C	I	I		E		¹⁵I		
T						M		¹⁶C	A	S	E	
¹⁷E				¹⁸I	E	¹⁹		T		A		
				E				²⁰				
²¹		²²S			²³R							
		C		²⁴E						²⁵A		
		A		D		²⁶				P		
²⁷N	A	N	D	G	A	T	E			P		
		N		E				²⁸		E		
		E								N		
	²⁹	R			³⁰					D		

40

Based on the **Dictionary of Computing**, third edition
ISBN 1-901659-04-6
Peter Collin Publishing Ltd

Communicative Crossword 2 — sheet B

Puzzles & Quizzes

This crossword is not complete: you have only half the words. The other half are on sheet A. Check that you know the words in your crossword. Then work with a partner who has sheet A to complete the two crosswords. Follow these three rules:

1. Speak only in English.
2. Don't say the word in the crossword.
3. Don't show your crossword to your partner.

"What's 1 across?"

→ across, ↓ down

¹D	I	²R	E	³C	T	⁴O	R	Y			
		A							⁵L	E	⁶D
⁷		M									O
								⁸			N
					⁹		¹⁰				G
¹¹I	N	K	¹²J	E	T						L
			U		¹³						E
¹⁴			S						¹⁵		
			T				¹⁶				
¹⁷E	F	F	I	C	¹⁸I	E	¹⁹N	T			
			F				I		²⁰P		
²¹B	U	²²S	Y		²³R	E	B	O	O	T	
R				²⁴			B		R		²⁵
A						²⁶S	L	O	T		
²⁷N						T	E				
C						A		²⁸L	O	S	E
H						C					
	²⁹T	R	E	E		³⁰K	E	Y	P	A	D

Based on the **Dictionary of Computing**, third edition

ISBN 1-901659-04-6

Peter Collin Publishing Ltd

41

Communicative crossword 3   sheet A

This crossword is not complete: you have only half the words. The other half are on sheet B. Check that you know the words in your crossword. Then work with a partner who has sheet B to complete the two crosswords. Follow these three rules:

1. Speak only in English
2. Don't say the word in the crossword.
3. Don't show your crossword to your partner.

"What's 1 across?"

→ across, ↓ down

¹A		²			³		⁴M		⁵R		⁶C	
L	■		■			■	A	■	O	■	L	
T	■		■			■	R	■	U	■	U	
E	■		■	⁷		■	K	■	T	■	S	
R	■		■			■	E	■	I	■	T	
N	■		■			■	⁸R	⁹A	N	¹⁰G	E	
A	■		■			■	¹¹P	E	E	R		
¹²T	O	O	L	B	O	X	■		■	T	■	
E	■		■		■		■		■		¹³	
■	■		¹⁴	■	¹⁵	¹⁶E					■	
¹⁷R	E	S	E	T	■	L		■		■		
E	■		■	■	¹⁸	I			■			
¹⁹S				■		M			■	²⁰		
I	■		■	■	²¹R	I	G	I	D	■		
²²D				■		N			■			
E	■		■	²³W	A	R	N	I	N	G		
N	■	■	■		■	T	■	■	■	■		
²⁴T	R	O	U	B	L	E	S	H	O	O	T	

42   Based on the **Dictionary of Computing**, third edition
ISBN 1-901659-04-6
Peter Collin Publishing Ltd

Communicative crossword 3   sheet B

This crossword is not complete: you have only half the words. The other half are on sheet A. Check that you know the words in your crossword. Then work with a partner who has sheet A to complete the two crosswords. Follow these three rules:

1. Speak only in English
2. Don't say the word in the crossword.
3. Don't show your crossword to your partner.

"What's one across?"

→ across, ↓ down

¹A	L	²P	H	A	³N	U	⁴M	E	⁵R	I	⁶C
		A			A						
T		R			N						
		T		⁷S	O	C	K	E	T		
		I			S						
N		T			E	⁸	⁹A		10	
		I			C		¹¹P			
12		O			O		P			
		N			N		L		¹³F
			¹⁴D		¹⁵D	¹⁶E	C	I	M	A	L
17			E						C		O
			L		¹⁸F	I	N	A	L		A
¹⁹S	O	L	I	D				T		²⁰X	T
			M		21			I			I
²²D	I	G	I	T				O			N
			T		23			N			G
24											

Based on the **Dictionary of Computing**, third edition
ISBN 1-901659-04-6
Peter Collin Publishing Ltd

43

Anagrams 2

Solve the anagrams by reading the clues and putting the letters of the words in order. Enter the solutions in the table to find the mystery phrase.

1. Machine that stores and processes data .... CEMOPRTU
2. To copy... ACEEILPRT
3. Scientific investigation to learn new facts about a field of study............................ ACEEHRRS
4. To keep within a limit.. CEIRRSTT
5. Fall, reduction.. ACDEEERS
6. To copy the behaviour of a system or device with another.................................. AEILMSTU
7. Always in the system .... DEEINRST
8. To keep in good working order .... AAIIMNNT
9. Unexplained fault in a system... EGILMNR
10. Graphical symbol used in programming as a sign for multiplication..................... AEIKRSST
11. Number of keys fixed together in some order, used to enter
    information into a computer .... ABDEKORY
12. Measure of the strength of a signal .... EIINNSTTY
13. One or more sectors on a hard disk that are used to store a file or part of a file........ CELRSTU
14. Loss or distortion of a signal .... ABEKPRU
15. To copy or behave like something else.. AEELMTU

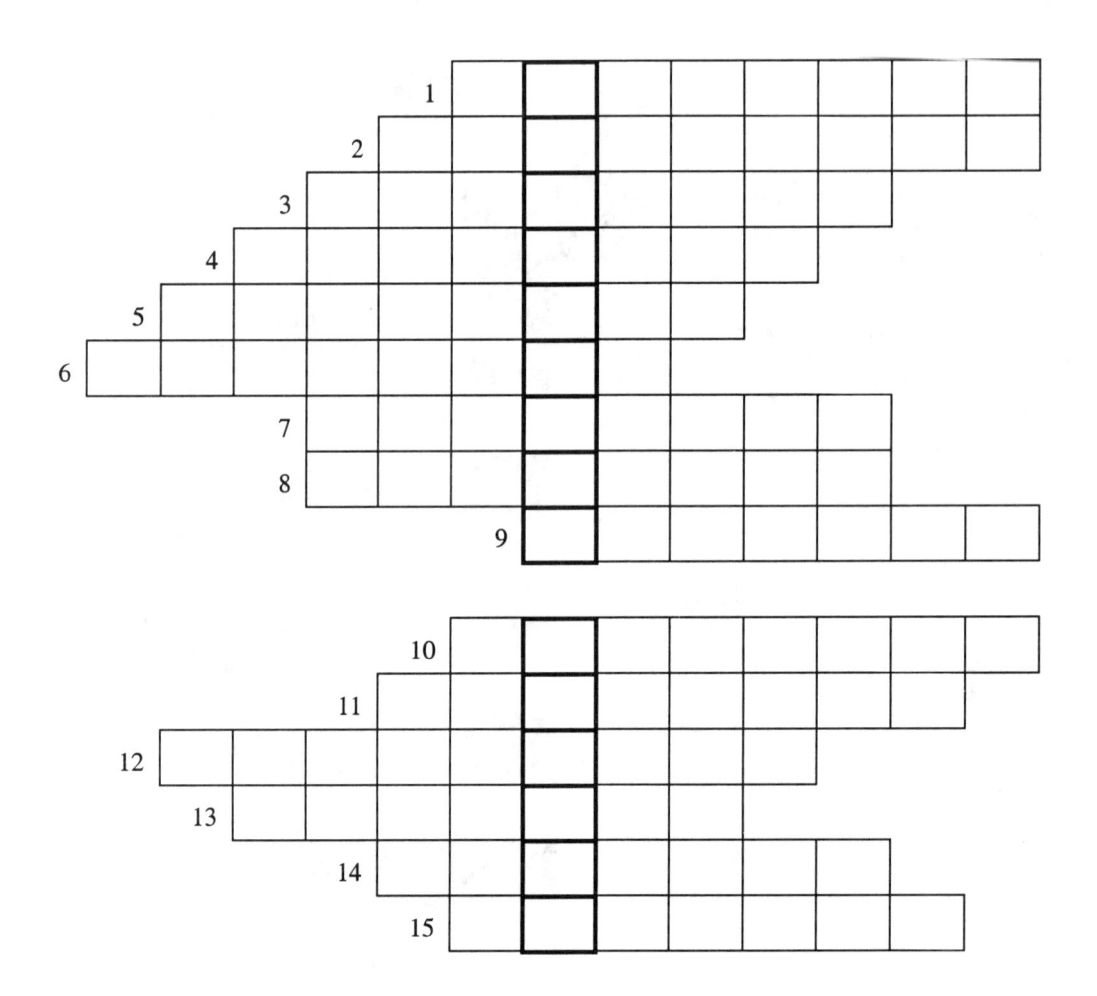

44 Based on the **Dictionary of Computing**, third edition
ISBN 1-901659-04-6
Peter Collin Publishing Ltd

Computing crossword

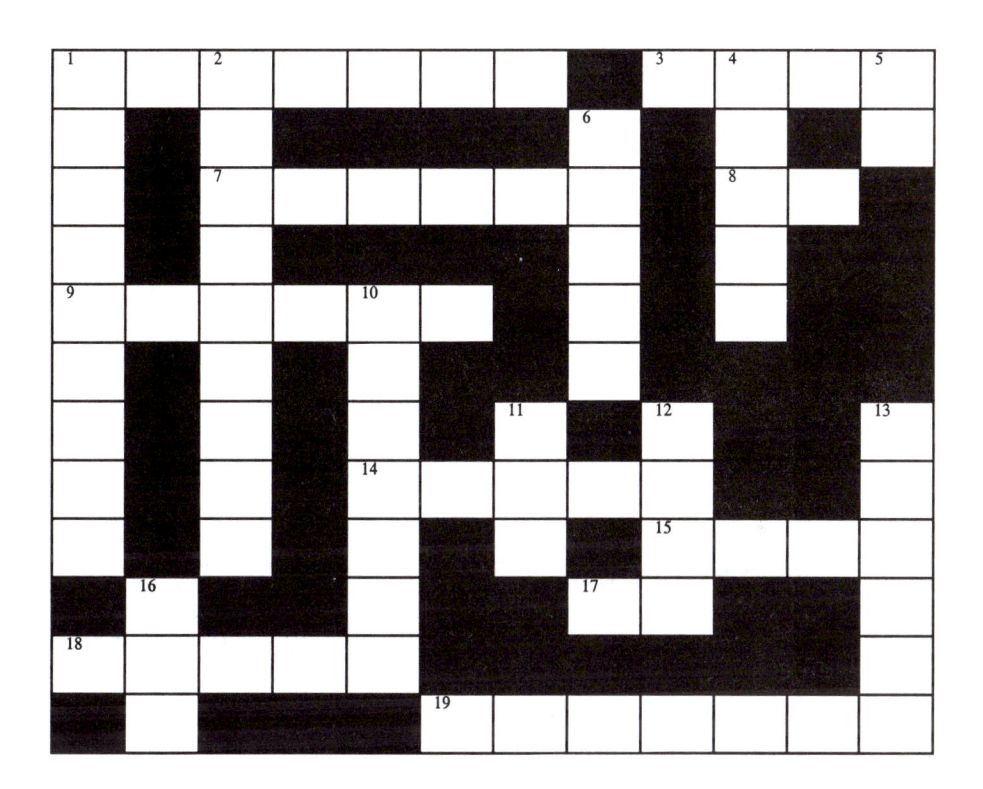

Across
1. Record of a user's name, password and rights to access a network or online system (7)
3. This dot-matrix printer is not a _____ printer; it only prints one line at a time (4)
7. Connected to and under the control of a central processor (6)
8. Abbreviation for internet protocol (2)
9. Main or most important device in a system (6)
14. To shut down access to a file (5)
15. To store data or a program on an auxiliary storage device (4)
17. Information technology (2)
18. Identification name given to a file, program or disk (5)
19. To link together two points in a circuit or communications network (7)

Down
1. Which works by itself, without being worked by an operator (9)
2. Interference between two communication cables or channels (9)
4. Different or not fitting the usual system (5)
5. Unit of measurement equal to half the width of an em (2)
6. Any physical material that can be used to store data (5)
10. To run or carry out a computer program or process (7)
11. Read only memory (3)
12. Action carried out on a device or program to establish whether it is working correctly, and if not, which component or instruction is not working (4)
13. To set something in advance (6)
16. Successful match or search of a database (3)

Based on the **Dictionary of Computing**, third edition
ISBN 1-901659-04-6
Peter Collin Publishing Ltd

Communicative crossword 4 — sheet A

This crossword is not complete: you have only half the words. The other half are on sheet B. Check that you know the words in your crossword. Then work with a partner who has sheet B to complete the two crosswords. Follow these three rules:

1. Speak only in English.
2. Don't say the word in the crossword.
3. Don't show your crossword to your partner.

"What's 1 across?"

→ across, ↓ down

¹H		²						³P			⁴
I								L			
⁵G	L	I	T		⁶C	H		U			
H					O		⁷S	I	G	⁸N	A	L
D					N					U		
⁹E					T					L		
N					I					L		
¹⁰S	E	¹¹T		N		12			C		
I					U					H		
¹³T					E					A		
Y						14				R		
S				¹⁵U					A		16
¹⁷T	R	A	N	S	F	E	¹⁸R		¹⁹C	A	D
O				E					T		
R				²⁰M	O	U	S	E			
²¹A								²²R			
G											
E		²³N	E	T	W	O	R	K	I	N	G

46 — Based on the **Dictionary of Computing**, third edition
ISBN 1-901659-04-6
Peter Collin Publishing Ltd

Communicative crossword 4   sheet B

This crossword is not complete: you have only half the words. The other half are on sheet A. Check that you know the words in your crossword. Then work with a partner who has sheet A to complete the two crosswords. Follow these three rules:

1. Speak only in English.
2. Don't say the word in the crossword.
3. Don't show your crossword to your partner.

"What's 1 across?"

→ across, ↓ down

¹H	O	²U	S	E	K	E	E	³P	I	N	⁴G
		N									O
⁵		I		⁶							A
		Q				⁷S			⁸		L
		U				P					
⁹E	L	E	C	T	R	I	C	A	L	L	Y
						L					
10		¹¹T			¹²L	A	T	C	H		
		R				A					
¹³T	R	A	C	E		G					
		N			¹⁴M	E	M	O	R	Y	
		S		15							¹⁶E
17		A				¹⁸R		19			D
		C				O					I
		T			20		U				T
²¹A	L	I	G	N			T		²²R	S	I
		O					E				N
		²³N					R				G

Based on the **Dictionary of Computing**, third edition
ISBN 1-901659-04-6
Peter Collin Publishing Ltd

Communicative crossword 5 — sheet A

This crossword is not complete: you have only half the words. The other half are on sheet B. Check that you know the words in your crossword. Then work with a partner who has sheet B to complete the two crosswords. Follow these three rules:

1. Speak only in English.
2. Don't say the word in the crossword.
3. Don't show your crossword to your partner.

"What's 1 across?"

→ across, ↓ down

¹		²			³		⁴	■	⁵		⁶
	■		■		■	⁷				■	
	■	⁸		⁹U		■		■		■	
¹⁰			■	N		■		■		■	
	■	¹¹E	N	A	B	L	I	N	G	■	
	■		■	L	■		■		■		
■	■	¹²		L					■		■
¹³			■	O	■		■		¹⁴		■
	■		■	¹⁵W	A	K	E	U	¹⁶P	■	
	■		■	A	■	■	■	■	O	■	
	■	¹⁷S	U	B	C	¹⁸L	A	S	S	■	
	■		■	L	■	O	■	■	T	■	■
¹⁹N	O	²⁰I	S	E	■	C	■	²¹S		² ²	
■	■	K	■	²³D	E	A	L	■	C	■	
²⁴P		B	■	I	■	T	■	²⁵D	R	A	G
²⁶U	N	S	I	G	N	E	D	■	I	■	■
S	■	■	■	I	■	■	■	²⁷U	P	■	■
²⁸H	I	G	H	T	E	C	H	■	²⁹T	A	B

48   Based on the **Dictionary of Computing**, third edition
ISBN 1-901659-04-6
Peter Collin Publishing Ltd

Communicative crossword 5   sheet B

This crossword is not complete: you have only half the words. The other half are on sheet A. Check that you know the words in your crossword. Then work with a partner who has sheet A to complete the two crosswords. Follow these three rules:

1. Speak only in English.
2. Don't say the word in the crossword.
3. Don't show your crossword to your

"What's 1 across?"

→ across, ↓ down

¹S	O	²F	T	W	³A	R	⁴E	■	⁵A	■	⁶T
Y	■	I	■	■	I	■	⁷L	A	S	E	R
M	■	⁸B	A	⁹U	D	■	I	■	S	■	A
¹⁰B	A	R	■	■	■	■	M	■	I	■	N
O	■	¹¹E	■	■	■	■	I	■	G	■	S
L	■	O	■	■	■	■	N	■	N	■	A
■	■	¹²P	U	L	L	■	A	■	■	■	C
¹³H	O	T	■	■	■	■	T	■	¹⁴I	■	T
I	■	I	■	¹⁵	■	■	E	■	¹⁶	■	I
D	■	C	■	■	■	■	■	■	■	■	O
D	■	¹⁷S	■	■	■	¹⁸	■	■	■	■	N
E	■	■	■	■	■	■	■	■	■	■	■
¹⁹N	■	²⁰	■	■	■	■	■	■	²¹S	E	²²T
■	■	■	■	²³	■	■	■	■	■	■	A
²⁴	■	■	■	■	■	■	²⁵	■	■	■	G
²⁶	■	■	■	■	■	■	■	■	■	■	■
■	■	■	■	■	■	■	²⁷	■	■	■	■
²⁸	■	■	■	■	■	■	■	²⁹	■	■	■

Based on the **Dictionary of Computing**, third edition
ISBN 1-901659-04-6
Peter Collin Publishing Ltd

Quiz

Can you answer these questions?

1. Exactly how many bytes are there in a megabyte?
   ....

2. Who invented the forerunner of today's digital computer?
   ....

3. Which company developed the first personal computer based on the Intel™ processor?
   ....

4. What does ISP stand for?
   ....

5. What is the name of the proprietary communications protocol that carries data over network hardware between two or more Apple Macintosh™ computers and peripherals?
   ....

6. When creating documents for the World Wide Web, do you use HTML or SGML?
   ....

7. What does it mean if your computer hangs?
   ....

8. Which company is the biggest developer and publisher of software for the PC and Macintosh?
   ....

9. What is the difference between a LAN and a WAN?
   ....

10. What is a browser?
    ....

11. In a computing context, what does 'handshaking' mean?
    ....

12. If you heard someone talking about a 'gooey', what would they be referring to?
    ....

Extension. Work with a partner and write a computer quiz. Make sure you know the answers. Then ask the questions to another pair of students.

Peter Collin Publishing

Vocabulary Record Sheet

WORD	CLASS	NOTES Translation or definition, example...

Based on the **Dictionary of Computing**, third edition
ISBN 1-901659-04-6
Peter Collin Publishing Ltd

Answer key

Word-building

Word association 1: missing links (p.1)
1. mouse 2. database 3. disk 4. printer
5. screen 6. file

Two-word expressions 1 (p.2)
1. parallel processing
2. integrated circuit
3. relational database
4. desktop publishing
5. artificial intelligence
6. optical fibre
7. preemptive multitasking
8. user-friendly
9. hard disk
10. clip-art
11. expanded memory
12. electronic mail
13. operating system
14. information technology
15. read only

Word formation: nouns (p.3)
1. There's a record of the user's new transaction in the database.
2. The installation of this system is easy.
3. There's slight screen flicker.
4. The launch of the new PC will be in January.
5. There was a system failure when I booted up this morning.
6. The factory has computer controlled production equipment *or* equipment for computer controlled production.
7. A maths co-processor is an enhancement of your system.
8. You'll have to make a comparison of the results of the two programs.
9. This is our storage system for client records.
10. Only privileged users have access to this information.
11. Recovery of data from a corrupted disk is sometimes possible.
12. Files retrieval is automatic.
13. Jack is responsible for system maintenance *or* maintenance of the system.
14. Something's wrong: there's no keyboard response.

Two-word expressions 2 (p.4)
1. machine code
2. flip-flop
3. floppy disk
4. clean machine
5. speech recognition
6. root directory
7. systems analysis
8. virus detector
9. laser printer
10. local bus
11. interactive video
12. graceful degradation
13. catastrophic error
14. baud rate
15. device driver

Plural formation (p.5)
1. viruses 2. expansion cards 3. appendices, 4. keys
5. asterisks 6. pixels 7. axes
8. directories 9. criteria 10. bureaux
11. formulae 12. fonts

Three-word expressions (p.6)
1. central processing unit
2. repetitive strain injury
3. graphical user interface
4. random access memory
5. bulletin board system
6. pull down menu
7. query by example
8. local area network
9. dynamic data exchange
10. dots per inch
11. optical character recognition
12. near letter quality

Word formation: adjectives (p.7)
1. compatible
2. confident
3. sophisticated
4. valid
5. electroluminescent
6. legible
7. different
8. efficient
9. capable
10. corrupt

Opposites 1: prefixes (p.8)
Exercise 1.

il-
1. illegal
2. illegible
3. illiterate

in-
1. inaccurate
2. inactive
3. incompatible
4. incorrect
5. indirect

un-
1. unauthorized
2. unformatted
3. unprotected
4. unjustified
5. unmodulated
6. undetected

Exercise 2.
1. unformatted 2. incorrect 3. incompatible 4. undetected 5. illegible 6. illiterate
7. inaccurate 8. unauthorized

Word formation: verbs (p.9)
Exercise 1.
1. alter 2. analyse 3. assemble 4. automate
5. communicate 6. compile 7. emulate
8. enhance 9. fluctuate 8. generate 9. install
10. instruct 11. interact 12. modify
13. multiply 14. prevent 15. process
16. program 17. recover 18. remove
19. retrieve 20. scan 21. store 22. use

Word association 2: mind maps (p.10)
1. caps
2. style sheet
3. clip-art
4. font
5. kerning
6. body
7. bit-mapped graphics
8. hi-res
9. image
10. heading
11. vector graphics
12. page preview
13. crop mark
14. hairline rule

Parts of Speech

Nouns (p.11)
1. database 2. password 3. modem 4. connector
5. field 6. model 7. procedure 8. utility 9. fault
10. plaintext 11. platform 12. virus

Adjectives (p.12)
1. re-chargeable 2. clean 3. common 4. concurrent
5. dedicated 6. normal 7. unformatted 8. faulty
9. crash-protected 10. electroluminescent
11. excessive 12. user-friendly 13. downloadable
14. redundant 15. unpopulated

Verbs 1 (p.13)
1. expand 2. halt 3. process 4. purge
5. undo 6. configure 7. save 8. recover
9. generate 10. simulate 11. simplify 12. run
13. install 14. paste 15. disconnect

Verbs 2 (p.14)
Exercise 1.
1.e 2.g 3.i 4.f 5.b 6.k 7.d 8.h 9.a 10.l
11.j 12.c

Exercise 2.
1. relayed 2. input 3. toggled 4. dumped
5. serviced 6. broadcast 7. monitors
8. emulate 9. overwritten 10. assigned
11. highlighted 12. launch

Verbs: past tense ~ regular verbs (p.15)
1. upgraded 2. designed 3. disabled
4. crashed 5. established 6. failed 7. viewed
8. received 9. updated 10. exported
11. demonstrated 12. checked 13. printed
14. contained 15. entered

Verbs: mixed tenses (p.16)
1. accessed 2. caches 3. computerized
4. conform 5. extract 6. identified
7. copying 8. lost 9. modified
10. running 11. shared 12. supersedes
13. carried 14. degaussed 15. housed

Phrasal verbs 1 (p.17)
1.i 2.d 3.k 4.j 5.e 6.a 7.f 8.n 9.m 10.h
11.b 12.g 13.l 14.c

Phrasal verbs 2 (p.18)
1. plugged in
2. log on
3. turns on
4. keyed in
5. back up
6. called up
7. shut down
8. broken down
9. turn off
10. log off
11. warming up
12. round down

Verbs: active/passive (p.19)
1. The syntax errors were automatically corrected by the compiler.
2. A balun was used to connect the coaxial cable to the twisted-pair circuit.
3. A maximum cable length of around 300m is supported by baseband local area networks.
4. Keyboarding costs were calculated on the basis of 5,500 keystrokes per hour.
5. The beta test software was tried out on as many different PCs as possible to try and find all the bugs.
6. A new brand of screen cleaner is being developed by this company.
7. Some defects in the equipment were found by the maintenance engineer.
8. The fault was traced to a faulty cable.
9. The copy flow is controlled by that device.
10. A metafile is used by the operating system to hold data that defines where each file is stored on the disk.

Adverbs (p.20)
1. We deal *directly* with the manufacturer, without using a wholesaler.
2. They were both *equally* responsible for the successful launch of the new system.
3. The text is *electronically* transmitted to an outside typesetter.
4. The files are arranged *alphabetically* under the customer's name.
5. Processing time is *approximately* 10% lower than during the previous quarter.
6. The sections of the program run *consecutively*.
7. A daisy wheel printer produces *fully* formed characters.
8. In spooling, the printer is acting *independently* of the keyboard.
9. The data was *incorrectly* keyboarded.
10. The paper has to be fed into the printer *manually*.

Prepositions (p.21)

1. The computer is a great aid to rapid processing in^large amounts of data. *of*
2. His background is of^the computer industry. *in*
3. Copy the files by^the hard drive, C:, to the floppy drive, A:. *from*
4. The cable has the wrong connector^this printer. *for*
5. The company is trying to improve the circulation^information between departments. *of*
6. The old data was contrasted at the latest information. *with*
7. The smoke at^the faulty machine quickly diffused through the building. *from*
8. The user cannot gain access^the confidential information in the file without a password. *to*
9. The glare from the screen makes to my eyes hurt.
10. The company has been illegally copying at copyright software.
11. The software is manufactured in this country after^licence. *under*
12. We had a new phone system installed in last week.

Pronunciation

Word stress (p.22)
Group A:
operate, processor, portable, memory, megabytes, graphical, interface, easily, generate
Group B:
italics, position, beginning, computer, transactions, recorded, efficient, statistics
Group C:
interact, represent, introduce

Present simple (p.23)
Group A:
looks, boots up, prints, corrects, outputs, detects, eliminates, checks, adapts, updates
Group B:
downloads, designs, programs, sends, receives, reads, conforms, installs, ensures
Group C:
browses, accesses, crashes, faxes

Past simple (p.24)
Group A:
introduced, launched, typed, released, checked, serviced, dumped, searched, networked
Group B:
received, found, closed down, opened, warned, programmed, moved, saved, reconfigured
Group C:
downloaded, selected, printed, deleted, recorded, eliminated, highlighted

Vocabulary in Context

Good advice (p.25)
1.o 2.n 3.g 4.d 5.a 6.m 7.c 8.e 9.k
10.j 11.p 12.f 13.i 14.b 15.h 16.l

Odd one out (p.26)

1. A **desktop** computer is fixed, the others are all portable machines.
2. A **modem** sends and receives information: the others are purely input devices.
3. **Soft** is the only one that is not a type of disk.
4. **DPI** (dots per inch) measures print density: the others measure speed.
5. The **motherboard** is the only one that is completely enclosed in the machine: the others can be seen from outside.
6. The **port** is the only one which can be seen from outside the computer.
7. A **model** is a version of a product: the others are types of port.
8. A **keyboard** only inputs information: the others can output information.
9. **File**: the others are user applications.
10. A **standalone** computer is the only one not connected to a network.
11. **Printout** is information on paper: the others are types of electronic memory.
12. A **window** is not part of the structure of a spreadsheet: the others are.
13. **Click** is an action performed with a mouse: the others are all styles of typeface.
14. A **screen** is the only one which is not a key on the keyboard.
15. **Message**: the others are all words which mean "mistake"
16. **Allocate** (the way in which a computer divides and assigns processing tasks) is the only one which does not mean "copy"
17. **Type** is the only one which is not something you do with a mouse
18. **QBE** (query by example) has no connection with the display system
19. **Scanner**: the others are types of printer
20. **Beep** is the sound a PC makes: the others are ways of eliminating data.

Opposites 2 (p.27)

Exercise 1.
1. anode - cathode
2. authorize - forbid
3. automatic - manual
4. backward - forward
5. boot up - shut down
6. character based - graphical
7. column - row
8. confirm - cancel
9. continue - interrupt
10. delete - restore
11. flexible - rigid
12. hardware - software
13. landscape - portrait
14. multiply - divide
15. open - close
16. add-on - built-in
17. simplex - duplex
18. simple - complicated
19. single - multiple
20. transmit - receive

Exercise 2.
1. software
2. forward
3. built-in
4. column
5. authorize

Abbreviations *(p.28)*

1. Bulletin Board System
2. Basic Input/Output System
3. bits per second
4. Computer-Assisted Design or Computer-Aided Design
5. Data Processing
6. dots per inch
7. File Allocation Table
8. High Memory Area
9. Intelligent Knowledge-Based System
10. Industry Standard Architecture
11. Information Technology
12. Local Area Network
13. Liquid Crystal Display
14. Million Instructions Per Second
15. Optical Character Recognition
16. Operating System
17. Personal Digital Assistant
18. Query by Example
19. Reduced Instruction Set Computer
20. Terminate and Stay Resident
21. Wide Area Network
22. Windows, Icons, Mouse, Pointers
23. Write Once, Read Many times
24. Word Processing
25. What You See Is What You Get

Telephone conversations *(p.29)*

Conversation A
1. PC Memory Mart. Can I help you?
2. Yes. Could you could give me some information about memory for Zell computers?
3. Certainly. Which model do you have?
4. The 333S.
5. And how much memory have you got in it at the moment?
6. Four megabytes.
7. There's a six meg module - that's an upgrade to ten.
8. That sounds alright. How much is it?
9. £240. Would you like to order it now?
10. No. No, I'd like to think about it. Thank you for the information.
11. Not at all. Goodbye.
12. Goodbye.

Conversation B
1. Viking Computers. Can I help you?
2. Yes. I've just bought one of your machines and I've got a problem.
3. I'm sorry to hear that. What kind of problem is it?
4. I can't boot up the system.
5. Do you get an error message when you try?
6. No - it just goes down a minute or two after starting.
7. I see. Can you give me your name and number? I'll get a technician to call you.
8. Yes. I'm Arthur Dent, D-E-N-T, and my number's 223 9898.
9. Someone will call you within an hour, Mr Dent.
10. Thank you.
11. Not at all. Goodbye.
12. Goodbye.

Operating systems *(p.30)*

Text 1. MS-DOS™
MS-DOS™ is operating system software developed by Microsoft that controls and co-ordinates the basic functions of your computer. If you are using Windows 95 or a later version of Windows, the functions of MS-DOS have been integrated. If you are using Windows 3.1x or do not have Windows, then you are relying on MS-DOS (or a similar product from IBM called PC-DOS) to control the computer.

Text 2. Windows™
Windows™ is a multitasking graphical user interface for the IBM PC developed by Microsoft Corp. that is designed to be easy to use. Windows™ uses icons to represent files and devices and can be controlled using a mouse, unlike MS-DOS which requires commands to be typed in.

Text 3. Windows 95™
Windows 95™ provides support for long filenames, an interface that's easier to use and better support for networks and the Internet. It does, however, require a faster processor and more memory to get good results - an absolute minimum of 8Mb and a fast 80486 are required.

Text 4. Windows 98™
Windows 98™ is an enhanced version of Microsoft's Windows 95™ that provides more communications and internet features and is easier to use and configure.

Instructions *(p.31)*

Instruction set A
Possible title: Formatting a disk using Windows 95™
1. Turn on the computer using the power switch on the front.
2. Turn on the monitor.
3. Put a new disk in the drive.
4. Double-click on the My Computer icon
5. Click on the 3-$^1/_2$ Floppy (A): icon.
6. Click on File in the menu bar and select Format from the pull-down menu.

Instruction set B
Possible title: Writing and printing a text in Word™
1. Click on Start, Programs and Microsoft Word™.
2. Type your text.
3. Click on File in the menu bar.
4. Select Print from the pull-down menu.
5. Click on OK in the dialogue box.
6. Collect your printout from the printer.

Instruction set C
Possible title: Fitting a new memory board in a printer
1. Turn off the computer and printer.
2. Disconnect the printer from the power supply and unplug the parallel interface cable.
3. Open the panel at the back.
4. Slide the old memory board out of the slot.
5. Push the new memory board in and replace the panel.
6. Reconnect the printer to the PC and to the power supply.

Memory (p.32)

1. (iii) (e)
2. (vi) (h)
3. (v) (b)
4. (vii) (c)
5. (ii) (d)
6. (viii) (g)
7. (iv) (a)
8. (i) (f)

Internet (p.33)

The internet is an international network that links together thousands of computers using telephone and cable links; these computers are called the servers and are rather like a local telephone exchange - individual users can then use a modem to connect to the server computer and so have access to the entire world network. A user can send electronic mail over the internet and transfer files and text from one computer in London to another in New York - all for the price of a local call to your nearest server.

The World Wide Web is an enhancement to the internet and provides a graphical front-end to the different databases and servers that are available. In order to connect to the internet, you'll need a modem and an account with a server - normally called an internet service provider (ISP) or point-of-presence provider - together with some special software. You'll be given a unique ID name that will (like your telephone number) identify you to any other user in the world together with an access telephone number that allows your modem to connect to their server. No one person or company controls the internet.

This and that (p.34)

1. bells and whistles
2. search and replace
3. cut and paste
4. wow and flutter
5. point and click
6. terminate and stay resident
7. drag and drop
8. hyphenation and justification
9. rows and columns
10. object linking and embedding
11. tilt and swivel

Slang (p.35)

1. comms
2. Big Blue
3. gremlin
4. vapourware
5. zap
6. smoke test
7. megs
8. kludge
9. glitch
10. cheapernet

Puzzles and Quizzes

Anagrams 1 (p.38)

1. READY
2. TABLE
3. INTERFACE
4. DIRECTORY
5. LANGUAGE
6. GARBAGE
7. IMPLEMENT
8. IDENTIFY
9. CLEAN
10. SIMULTANEOUS
11. MACHINE
12. CUSTOMIZE
13. QUINTET
14. RANDOM
15. INSTALL

Word search (p.39)

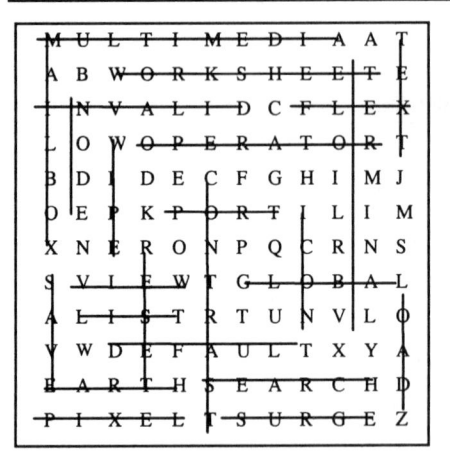

MULTIMEDIA words found: MULTIMEDIA, WORKSHEET, INVALID, FLEX, LOW, OPERATOR, PORT, VIEW, GLOBAL, LIST, DEFAULT, EARTH, SEARCH, PIXEL, SURGE

Anagrams 2 (p.44)

1. COMPUTER
2. REPLICATE
3. RESEARCH
4. RESTRICT
5. DECREASE
6. SIMULATE
7. RESIDENT
8. MAINTAIN
9. GREMLIN
10. ASTERISK
11. KEYBOARD
12. INTENSITY
13. CLUSTER
14. BREAKUP
15. EMULATE

Computing crossword *(p.45)*

Across
1. account,   3. page,   7. online,   8. IP,
9. master,   14. close,   15. save,   17. IT,   18. title,
19. connect

Down
1. automatic,   2. crosstalk,   4. alien,
5. en,   6. media,   10. execute,   11. ROM,
12. down   13. preset,   16. hit,   17. test

Quiz *(p.50)*

1. 1,048,576
2. Charles Babbage
3. IBM (International Business Machines)
4. internet service provider
5. AppleTalk™
6. HTML
7. It enters an endless loop and will not respond to further instruction.
8. Microsoft™
9. A LAN (local area network) is a network in which various terminals and equipment are all within a short distance of one another and can be interconnected by cables, whereas in a WAN (wide area network) the various terminals are far apart and linked by radio, satellite and cable.
10. A software program that is used to navigate through pages stored on the internet.
11. It refers to the standardised signals between two devices to make sure that the system is working correctly, the equipment is compatible and data transfer is correct.
12. A Graphical User Interface (GUI - pronounced 'gooey'), which is an interface between an operating system or program and the user; it uses graphics or icons to represent functions or files and allow the software to be controlled more easily.

SPECIALIST DICTIONARIES

DICTIONARY OF COMPUTING

The ideal companion to this workbook. An up-to-date, comprehensive dictionary that includes over 10,000 entries and provides clear definitions of the terms used in networking, hardware, software, PCs, mini computers and mainframes, the Internet, multimedia, and programming.
- part of speech, grammar notes and encyclopaedic comments
- quotations from newspapers and magazines show how terms are used
- example sentences and quotations show usage

ISBN 1-901659-04-6   390pages paperback   £9.95 / US$15.95

ENGLISH DICTIONARY FOR STUDENTS

A new, general English dictionary for students. Provides an up-to-date, clear definitions of over 25,000 entries. The terms cover British, American and international English; the word list has been selected to be useful for students taking Cambridge First Certificate, SAT-1 or other similar examinations. Each entry include phonetic pronunciation, part of speech, clear definition, example sentences to show use.

ISBN 1-901659-06-2   720pages paperback   £9.95 / US$15.95

Visit our web site: www.pcp.co.uk   for details on our complete list, demo software and resource materials.
Or use the form below to request further information.

English Dictionaries

Accounting	0-948549-27-0
Aeronautical Terms	1-901659-10-0
Agriculture, 2nd ed	0-948549-78-5
American Business	0-948549-11-4
Automobile Engineering	0-948549-66-1
Banking & Finance, 2nd ed	1-901659-30-5
Business, 2nd ed	0-948549-51-3
Computing, 3rd ed	1-901659-04-6
Ecology & Environment, 3rd ed	0-948549-74-2
English Dictionary for Students	1-901659-06-2
Government & Politics, 2nd ed	0-948549-89-0
Hotel, Tourism, Catering Management	0-948549-40-8
Human Resources & Personnel, 2nd ed	0-948549-79-3
Information Technology, 2nd ed	0-948549-88-2
Law, 2nd ed	0-948549-33-5
Library & Information Management	0-948549-68-8
Marketing, 2nd ed	0-948549-73-4
Medicine, 2nd ed	0-948549-36-X
Military Terms	1-901659-24-0
Printing & Publishing, 2nd ed	0-948549-99-8
Science & Technology	0-948549-67-X

Vocabulary Workbooks

Banking & Finance	0-948549-96-3
Business, 2nd ed	1-901659-27-5
Computing, 2nd ed	1-901659-28-3
Colloquial English	0-948549-97-1
English	1-901659-11-9
Hotels, Tourism, Catering	0-948549-75-0
Law, 2nd ed	1-901659-21-6
Medicine	0-948549-59-9

Professional/General

Astronomy	0-948549-43-2
Economics	0-948549-91-2
Multimedia, 2nd ed	1-901659-01-1
PC & the Internet, 2nd ed	1-901659-12-7
Bradford Crossword Solver, 3rd ed	1-901659-03-8
Bradford Crossword Key	1-901659-40-2

Bilingual Dictionaries

French-English/English-French Dictionaries
German-English/English-German Dictionaries
Spanish-English/English-Spanish Dictionaries

Name: ....
Address: ....
....
....
....Postcode: ....Country: ....

PCP

Peter Collin Publishing Ltd
1 Cambridge Road
Teddington, TW11 8DT - UK
tel: +44 181 943 3386   fax: +44 181 943 1673   email: info@pcp.co.uk
web site: www.pcp.co.uk